Introduction to AutoCAD 2009

WITHDRAWN

Introduction to AutoCAD 2009

2D and 3D Design

First edition

Alf Yarwood

AMSTERDAM • BOSTON • HEIDELBERG • LONDON • NEW YORK • OXFORD
PARIS • SAN DIEGO • SAN FRANCISCO • SINGAPORE • SYDNEY • TOKYO

Newnes is an imprint of Elsevier

Newnes is an imprint of Elsevier

Linacre House, Jordan Hill, Oxford OX2 8DP, UK
30 Corporate Drive, Suite 400, Burlington, MA 01803, USA

First edition 2008

British Library Cataloguing in Publication Data
A catalogue record for this book is available from the British Library

Library of Congress Cataloging-in-Publication Data
A catalog record for this book is availabe from the Library of Congress

ISBN: 978-0-7506-8983-0

For information on all Newnes publications
visit our web site at books.elsevier.com

Typeset by Charon Tec Ltd., A Macmillan Company. (www.macmillansolutions.com)

Printed and bound by MKT, Slovenia

08 09 10   10 9 8 7 6 5 4 3 2 1

Contents

Preface

The purpose of writing this book is to produce a text suitable for students in Further and/or Higher Education who are required to learn how to use the CAD software package AutoCAD® 2009. Students taking examinations based on computer-aided design will find the contents of the book of great assistance. The book is also suitable for those in industry wishing to learn how to construct technical drawings with the aid of AutoCAD 2009 and those who, having used previous releases of AutoCAD, wish to update their skills to AutoCAD 2009.

The chapters in Part 1: 2D Design, dealing with two-dimensional drawing, will also be suitable for those wishing to learn how to use AutoCAD LT 2009, the two-dimensional (2D) version of this latest release of AutoCAD.

Many readers using previous releases of AutoCAD will find the book's contents largely suitable for use with those versions, although AutoCAD 2009 has considerable enhancements over previous releases (some of which are mentioned in Chapter 21).

The contents of this book are basically a graded course of work, consisting of chapters giving explanations and examples of methods of constructions, followed by exercises which allow the reader to practise what has been learned in each chapter. The first 11 chapters are concerned with constructing technical drawing in two dimensions. These are followed by chapters detailing the construction of three-dimensional (3D) solid drawings and rendering them. The two final chapters describe the Internet tools of AutoCAD 2009 and the place of AutoCAD in the design process. The book finishes with three appendices: printing and plotting; a list of tools with their abbreviations; and a list of some of the set variables upon which AutoCAD 2009 is based.

AutoCAD 2009 is very complex computer-aided design (CAD) software package. A book of this size cannot possibly cover the complexities of all the methods for constructing 2D and 3D drawings available when working with AutoCAD 2009. However, it is hoped that by the time the reader has worked through the contents of the book, he/she will be sufficiently skilled in methods of producing drawing with the software to be able to go on to more advanced constructions with its use and will have gained an interest in the more advanced possibilities available when using AutoCAD.

Alf Yarwood

Registered Trademarks

Autodesk® and AutoCAD® are registered in the US Patent and Trademark Office by Autodesk Inc.

Windows® is a registered trademark of the Microsoft Corporation.

Alf Yarwood is an Autodesk authorized author and a member of the Autodesk Developer Network.

Part 1

2D Design

Introducing AutoCAD 2009

AIM OF THIS CHAPTER

The aim of this chapter is to introduce features of the AutoCAD 2009 window and methods of operating AutoCAD 2009.

Opening AutoCAD 2009

Fig. 1.1 The **AutoCAD 2009** shortcut icon on the Windows desktop

AutoCAD 2009 is designed to work in a Windows operating system. In general, to open AutoCAD 2009, either *double-click* on the **AutoCAD 2009** shortcut in the Windows desktop (Fig. 1.1), or *right-click* on the icon, followed by a *left-click* on **Open** in the menu which then appears (Fig. 1.2).

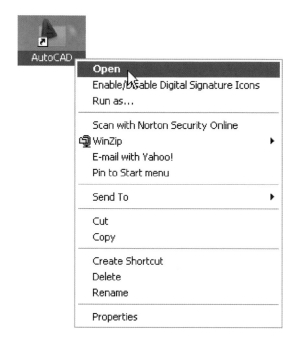

Fig. 1.2 The *right-click* menu which appears from the shortcut icon

When working in education or in industry, computers may be configured to allow other methods of opening AutoCAD, such as a list appearing on the computer in use when the computer is switched on, from which the operator can select the program he/she wishes to use.

When AutoCAD 2009 is opened a window appears, the window depending upon whether a **3D Modeling**, **Classic AutoCAD** or a **2D Drafting & Annotation** workspace has been set as the **QNEW** (see p. 276). In this example the **2D Drafting & Annotation** workspace is shown and includes the **Ribbon** with **Tool panels** (Fig. 1.3). This **2D Drafting & Annotation** workspace shows the following details:

- **Ribbon**: which includes tabs, each of which when *clicked* will bring a set of panels containing tool icons. Further tool panels can be seen by *clicking* the appropriate tab.
- **Menu Browser** icon: A *left-click* on the arrow to the right of the **A** symbol at the top-left-hand corner of the AutoCAD 2009 window causes the **Menu Browser** menu to appear (Fig. 1.4).

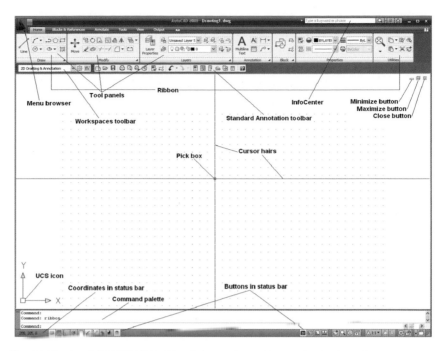

Fig. 1.3　The AutoCAD 2009 **2D Drafting & Annotation** workspace showing its various parts

Fig. 1.4　The **Menu Browser** menu

- **Workspaces Switching** menu: appears with a *click* on the **Workshop Switching** button in the status bar (Fig. 1.5).
- **Command palette**: can be *dragged* from its position at the bottom of the AutoCAD window into the AutoCAD drawing area, when it can be seen to be a palette (Fig. 1.6). As with all palettes, an **Auto-hide** icon and a *right-click* menu is included:
- **Panels**: each shows tools appropriate to the panel. Taking the **Home/ Draw** panel as an example, Fig. 1.7 shows that a *click* on one of the tool icons in the panel brings a tooltip on screen showing details of how the

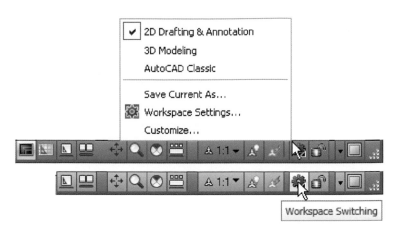

Fig. 1.5 The **Workspace Switching** dialog appearing when the **Workspace Switching** button is *clicked*

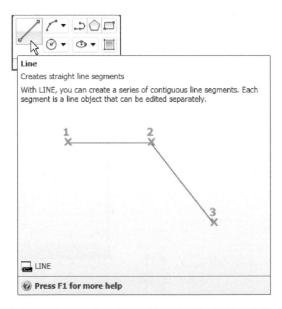

Fig. 1.6 The command palette when *dragged* from its position as the bottom of the AutoCAD window

Fig. 1.7 The descriptive tooltip appearing with a *click* on the **Line** tool icon in the **Home/Draw** panel

tool can be used. Other tool icons have a pop-up menu as a tooltip. In the example given in Fig. 1.8, a *click* on the **Circle** tool icon will show a tooltip. A *click* on the arrow to the right of the tool icon and a flyout appears showing the construction method options available for the tool.

- **Standard Annotation toolbar**: One of the toolbars just below the ribbon includes the **Open…** icon (Fig. 1.9). A *click* on the icon will bring the **Select File** dialog on screen.

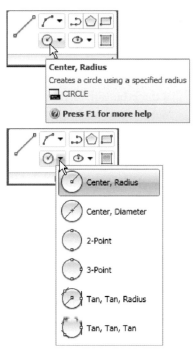

Fig. 1.8 The tooltip for the **Circle** tool and its pop-up menu

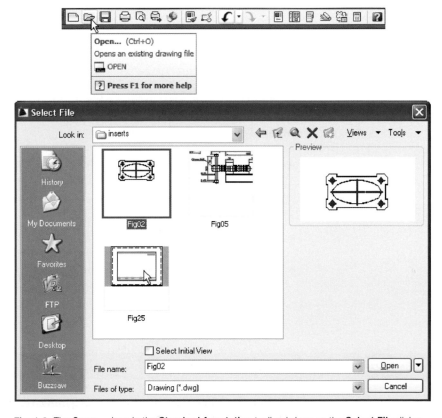

Fig. 1.9 The **Open…** icon in the **Standard Annotation** toolbar brings up the **Select File** dialog

The mouse as a digitizer

Many operators working in AutoCAD will use a two-button mouse as a digitizer. There are other digitizers which may be used – picks with tablets, a three-button mouse etc. Figure 1.10 shows a mouse which has two buttons and a wheel.

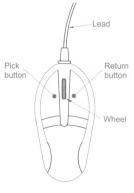

Fig. 1.10 A two-button mouse

To operate this mouse pressing the **Pick** button is a *left-click*. Pressing the **Return** button is a *right-click*. Pressing the **Return** button usually, but not always, has the same result as pressing the **Enter** key of the keyboard.

When the **wheel** is pressed, drawings in the AutoCAD screen can be panned by moving the mouse. Moving the wheel forwards enlarges (zooms in) the drawing on screen. Move the wheel backwards and a drawing reduces in size.

The pick box at the intersection of the cursor hairs moves with the cursor hairs in response to movements of the mouse. The AutoCAD window as shown in Fig. 1.3 includes cursor hairs which stretch across the drawing in both horizontal and vertical directions. Some operators prefer cursors hairs to be shorter. The length of the cursor hairs can be adjusted in the **Display** sub-menu of the **Options** dialog (p. 12).

Palettes

A palette has already been shown – the **Command** palette. Two palettes which may be frequently used are the **DesignCenter** palette and the **Properties** palette. These can be called to screen from icons in the **Standard Annotation** toolbar.

- **DesignCenter** palette: Fig. 1.11 shows the **DesignCenter** palette with the **Block** drawings of electronics circuit symbols from an AutoCAD directory **DesignCenter** from which the file **Fasteners – Metric.dwg** has been selected.
- **Properties** palette: Fig. 1.12 shows the **Properties** palette, called from the **Standard Annotation** toolbar, in which the general and geometrical features of a selected line are shown. The line can be changed by the *entering* of new figures in parts of the palette.

Tool palettes

Click on **Tool Palettes Window** in the **Standard Annotation** toolbar and the **Tool Palettes – All Palettes** palette appears (Fig. 1.13). *Right-click*

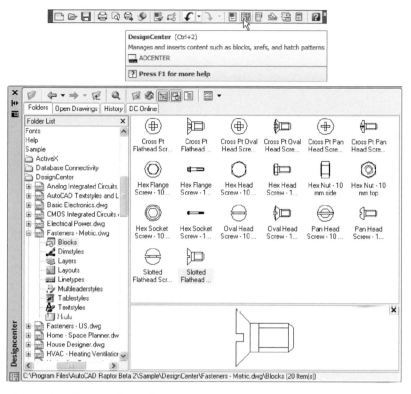

Fig. 1.11 A *left-click* on the **DesignCenter** icon brings the **DesignCenter** palette to screen

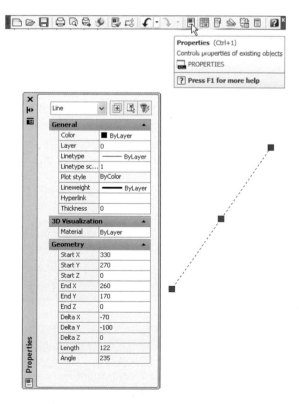

Fig. 1.12 The **Properties** palette

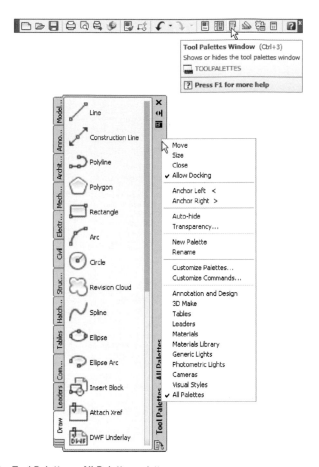

Fig. 1.13 The **Tool Palettes – All Palettes** palette

in the title bar of the palette and a pop-up menu appears. *Click* on a name in the menu and the selected palette appears together with those panels already selected from the pop-up list. The palettes can be reduced in size by *dragging* at corners or edges, or hidden by *clicking* on the **Auto-hide** icon, or moved by *dragging* on the **Move** icon. The palette can also be *docked* against either side of the AutoCAD window.

Note

Throughout this book tools will often be shown as selected from the panels. It will be seen in Chapter 3 that tools can be 'called' in a variety of ways, but tools will frequently be shown selected from tool panels although other methods will also be shown on occasion.

Dialogs

Dialogs are an important feature of AutoCAD 2009. Settings can be made in many of the dialogs, files can be saved and opened and changes can be made to variables.

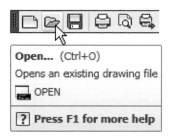

Fig. 1.14 Opening the **Select File** dialog from the **Open** icon in the **Standard Annotation** toolbar

Examples of dialogs are shown in Figs 1.15 and 1.16. The first example is taken from the **Select File** dialog (Fig. 1.15), opened with a *click* on **Open…** in the **Quick Access Toolbar** (Fig. 1.14). The second example shows part of the **Options** dialog (Fig. 1.16) in which many settings can be made to allow operators the choice of their methods of constructing drawings. The **Options** dialog can be opened with a *click* on **Options…** in the *right-click* dialog opened in the command palette.

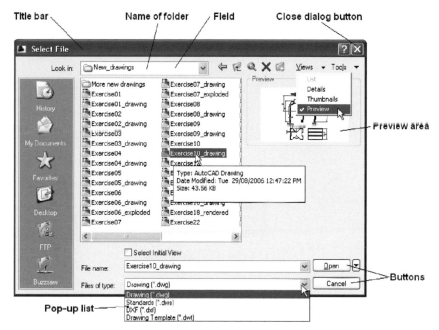

Fig. 1.15 The **Select File** dialog

Note the following parts in the dialog, many of which are common to other AutoCAD dialogs:

- **Title bar**: showing the name of the dialog.
- **Close dialog button**: common to other dialogs.
- **Pop-up list**: a *left-click* on the arrow to the right of the field brings down a pop-up list listing selections available in the dialog.
- **Buttons**: a *click* on the **Open** button brings the selected drawing on screen. A *click* on the **Cancel** button, closes the dialog.
- **Preview** area: available in some dialogs – shows a miniature of the selected drawing or other feature, partly shown in Fig. 1.15.

Note the following in the **Options** dialog:

- **Tabs**: a *click* on any of the tabs in the dialog brings a sub-dialog on screen.
- **Check boxes**: a tick appearing in a check box indicates that the function described against the box is on. No tick and the function is off. A *click* in a check box toggles between the feature being off or on.

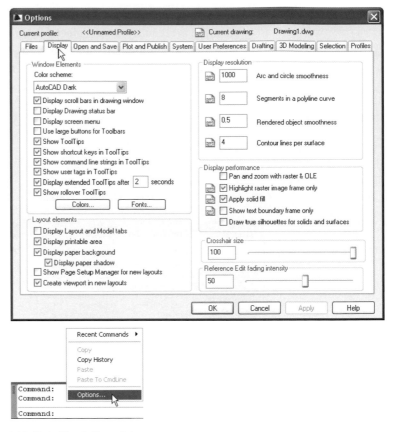

Fig. 1.16 Part of the **Options** dialog

- **Radio buttons**: a black dot in a radio button indicates the feature described is on. No dot and the feature is off.
- **Slider**: a slider pointer can be *dragged* to change sizes of the feature controlled by the slider.

Buttons at the left-hand end of the status bar

A number of buttons at the left-hand end of the status bar can be used for toggling (turning on/off) various functions when operating within AutoCAD 2009 (Fig. 1.17). A *click* on a button turns that function on; if it is off, a *click* on a button turns the function back on. Similar results can be obtained by using function keys of the computer keyboard (keys **F1** to **F10**).

- **Snap Mode**, also toggled using the **F9** key: when snap on, the cursor under mouse control can only be moved in jumps from one snap point to another (see also p. 71).
- **Grid Display**, also toggled using the **F7** key: when set on a series of grid points appears in the drawing area.

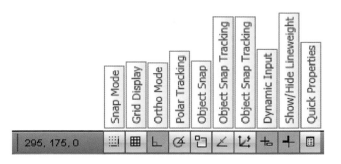

Fig. 1.17　The buttons at the left-hand end of the status bar

- **Ortho Mode**, also toggled using the **F8** key: when set on, lines etc., can only be drawn vertically or horizontally.
- **Polar Tracking**, also toggled using the **F10** key: when set on, a small tip appears showing the direction and length of lines etc., in degrees and units.
- **Object Snap**, also toggled using the **F3** key: when set on an OSnap icon appears at the cursor pick box (see also p. 71).
- **Object Snap Tracking**: when set on, lines etc., can be drawn at exact coordinate points and precise angles.
- **Allow/Disallow Dynamic UCS**: also toggled by the **F6** key. Used when constructing 3D solid models.
- **Dynamic Input**: also toggled by **F12**. When set on, the **x,y** coordinates and prompts show when the cursor hairs are moved.
- **Show/Hide Lineweight**: when set on, lineweights show on screen. When set off, lineweights only show in plotted/printed drawings.
- **Quick Properties**: a *right-click* brings up a pop-up menu, from which a *click* on **Settings…** causes the **Drafting Settings** dialog to appear.

Note

When constructing drawings in AutoCAD 2009 it is advisable to toggle between **Snap**, **Ortho**, **OSnap** and the other functions in order to make constructing easier.

Buttons at the right-hand end of the status bar

Another set of buttons at the right-hand end of the status bar are shown in Fig. 1.18. The uses of some of these will become apparent when reading future pages of this book. A *click* on the downward-facing arrow near the right-hand end of this set of buttons brings up the **Application Status Bar Menu** (Fig. 1.19) from which the buttons in the status bar can be set on or off.

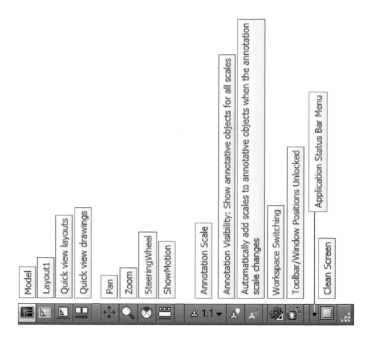

Fig. 1.18 The buttons at the right-hand end of the status bar

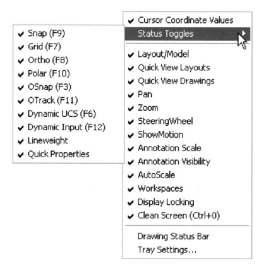

Fig. 1.19 The **Application Status Bar Menu**

The AutoCAD coordinate system

In the AutoCAD 2D coordinate system, units are measured horizontally in terms of X and vertically in terms of Y. A 2D point can be determined in terms of X,Y (in this book referred to as *x,y*). *x,y* = 0,0 is the **origin** of the system. The coordinate point *x,y* = 100,50 is 100 units to the right of the origin and 50 units above the origin. The point *x,y* = −100, −50 is 100 units to the left of the origin and 50 points below the origin. Figure 1.20 shows some 2D coordinate points in the AutoCAD window.

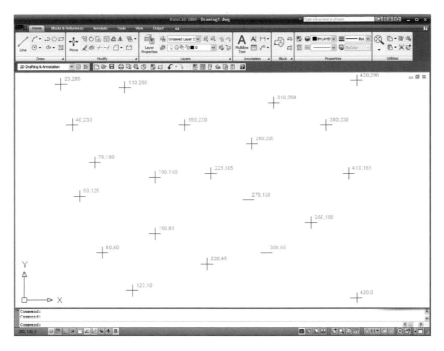

Fig. 1.20　The 2D coordinate points in the AutoCAD coordinate system

3D coordinates include a third coordinate (Z), in which positive Z units are towards the operator as if coming out of the monitor screen and negative Z units going away from the operator as if towards the interior of the screen. 3D coordinates are stated in terms of x,y,z . $x,y,z = 100,50,50$ is 100 units to the right of the origin, 50 units above the origin and 50 units towards the operator. A 3D model drawing as if resting on the surface of a monitor is shown in Fig. 1.21.

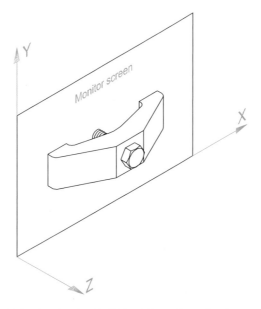

Fig. 1.21　A 3D model drawing showing the X, Y and Z coordinate directions

Drawing templates

Drawing templates are files with an extension **.dwt**. Templates are files which have been saved with predetermined settings – such as **Grid** spacing, **Snap** spacing etc. Templates can be opened from the **Select template** dialog (Fig. 1.22) called by *clicking* the **New...** icon at the top-left-hand corner of the AutoCAD window. An example of a template file being opened is shown in Fig. 1.22. In this example the template will be opened in Paper Space and is complete with a title block and borders.

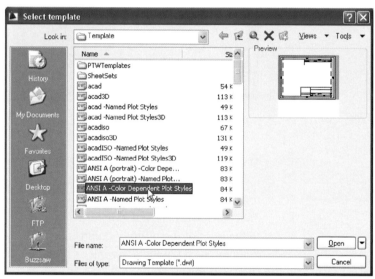

Fig. 1.22   A template selected for opening in the **Select template** dialog

When AutoCAD 2009 is used in European countries and opened, the **acadiso.dwt** template is the one most likely to appear on screen. In Part 1 of this book (2D Design), drawings will usually be constructed in an adaptation of the **acadiso.dwt** template. Adapt this template as follows:

1. In the command palette *enter* (type) **grid** followed by a *right-click* (or pressing the **Enter** key). Then *enter* **10** in response to the prompt which appears, followed by a *right-click* (Fig. 1.23).

```
Command: grid
Specify grid spacing(X) or [ON/OFF/Snap/Major/aDaptive/Limits/Follow/Aspect]
<10>:
Command:
```

Fig. 1.23   Setting **Grid** to **10**

2. In the command palette *enter* **snap** followed by *right-click*. Then *enter* **5** followed by a *right-click* (Fig. 1.24).
3. In the command palette *enter* **limits**, followed by a *right-click*. *Right-click* again. Then *enter* **420,297** and *right-click* (Fig. 1.25).

```
Command: snap
Specify snap spacing or [ON/OFF/Aspect/Style/Type] <0>: 5
Command:
```

Fig. 1.24 Setting **Snap** to **5**

```
Command: limits
Reset Model space limits:
Specify lower left corner or [ON/OFF] <0,0>:
Specify upper right corner <12,9>: 420,297
Command:
```

Fig. 1.25 Setting **Limits** to **420,297**

4. In the command palette *enter* **zoom** and *right-click*. Then in response to the line of prompts which appears *enter* **a** (for All) and *right-click* (Fig. 1.26).

```
Command: z
ZOOM
Specify corner of window, enter a scale factor (nX or nXP), or
[All/Center/Dynamic/Extents/Previous/Scale/Window/Object] <real time>: a
Regenerating model.
Command:
```

Fig. 1.26 **Zooming** to **All**

5. In the command palette *enter* **units** and *right-click*. The **Drawing Units** dialog appears (Fig. 1.27). In the **Precision** pop-up list of the **Length** area of the dialog, *click* on **0** and then *click* the **OK** button. Note the change in the coordinate units showing in the status bar.

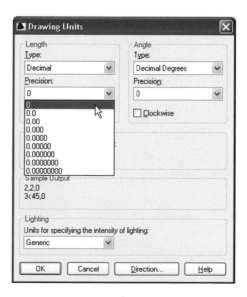

Fig. 1.27 Setting **Units** to **0**

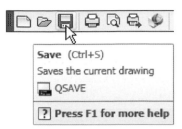

Fig. 1.28 *Click* **Save**

6. *Click* the **Save** icon in the **Standard Annotation** toolbar (Fig. 1.28). The **Save Drawing As** dialog appears. In the **Files of type** pop-up list select **AutoCAD Drawing Template (*.dwt)**. The templates already in AutoCAD are displayed in the dialog. *Click* on **acadiso.dwt**, followed by another *click* on the **Save** button.

Note

1. Now when AutoCAD is opened the template saved as **acadiso.dwt** automatically loads with **Grid** set to **10**, **Snap** set to **5**, **Limits** set to **420,297** (size of an A3 sheet in millimetres) and with the drawing area zoomed to these limits, with **Units** set to **0**.

2. However, if there are multiple users by the computer, it is advisable to save your template to another file name – e.g. **my_template.dwt**.

3. Other features will be added to the template in future chapters.

Method of showing entries in the command palette

Throughout the book, when a tool is 'called', usually by a *click* on a tool icon in a panel – in this example *picking* the **Navigation/Zoom** tool – the command palette will show as follows:

```
Command: _zoom right-click
Specify corner of window, enter a scale factor (nX
   or nXP), or [All/Center/Dynamic/Extents/
   Previous/Scale/Window/Object]<;real time>:
   enter a (All) right-click
Regenerating model.
Command:
```

Note: In later examples this may be shortened to:

```
Command: _zoom
[prompts]: enter a right-click
Command:
```

Note

1. In the above *enter* means type the given letter, word or words at the **Command:** prompt.

2. *Right-click* means press the **Return** (right) button of the mouse or press the **Return** key of the keyboard.

Tools and tool icons

In AutoCAD 2009, tools are shown as icons in panels or in toolbars. When the cursor is placed over a tool icon a description shows with the name of the tool as shown and an explanation in diagram form as in the example given in Fig. 1.7 (p. 6).

If a small arrow is included at the right-hand side of a tool icon, when the cursor is placed over the icon and the *pick* button of the mouse depressed and held, a flyout appears which includes other features. An example is given in Fig. 1.8 (p. 7).

Another AutoCAD workspace

Click **Workspace Switching** icon at the bottom-right-hand corner of the AutoCAD window (Fig. 1.29). In the menu which appears *click* the

Fig. 1.29 Selecting **AutoCAD Classic** from the **Workspace Switching** menu

box at the left of **AutoCAD Classic** (Fig. 1.29). The **AutoCAD Classic** workspace appears (Fig. 1.30). This includes the **Draw** toolbar *docked* against the left-hand side of the window and the **Modify** toolbar *docked* against the right-hand side of the window with other toolbars *docked* against the top of the window. The menu bar is included in the example given in Fig. 1.30.

Other workspaces can be designed as the operator wishes. One in particular which may appeal to some operators is to *click* the **Clear Screen** icon at the bottom-right corner of the AutoCAD window (Fig. 1.31). This example shows a clear screen window from the **AutoCAD Classic** workspace. This allows more working space.

The **Ribbon**

The **Ribbon** contains groups of panels placed, by default, at the top of the AutoCAD 2009 window. There are, by default, 6 groups of panels called

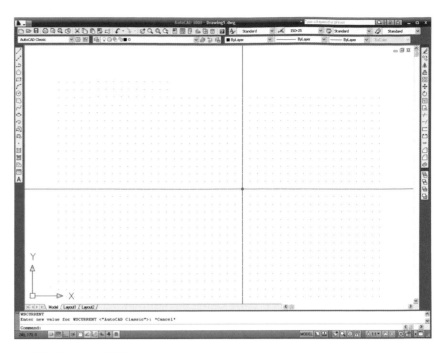

Fig. 1.30 The **AutoCAD Classic** window

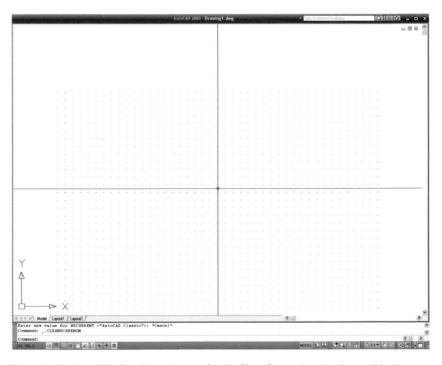

Fig. 1.31 The **AutoCAD Classic** workspace after the **Clear Screen** icon has been *clicked*

from **tabs** above the panels – **Home**, **Blocks & References**, **Annotate**, **Tools**, **View** and **Output**. Other groups can be added if wished by using the **Custom User Interface** dialog.

If a small arrow is showing in the bottom-right-hand corner of a panel, a *click* on the arrow brings down a flyout showing additional tool icons

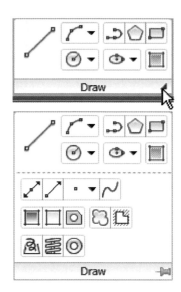

Fig. 1.32 The **Draw** panel and its pop-up

not included in the panel. As an example Fig. 1.32 shows the flyout from the **Draw** panel.

A *click* on **Undock** in the *right-click* menu from the arrow in the **Mimimize to panel titles** and the **Ribbon** becomes a palette (Fig. 1.33). A *right-click* menu shows that the palette can now be *docked* against either side of the window. Some operators will prefer selecting tools from a *docked* **Ribbon** because of the larger workspace obtained when it is so *docked*.

A *click* on the **Minimize to panel titles** followed by another *click* on **Mimimize to tabs** causes the **Ribbon** to disappear, leaving only the tabs showing, but each panel can then be selected from the panel tabs. Figure 1.34 shows the AutoCAD window with the **Ribbon** minimized to tabs and with the **Home** tab selected and with **Line** selected from the **Draw** panel. When the **Line** tool comes into action the **Home** panels disappear again, leaving a larger drawing area than before the panels were minimized.

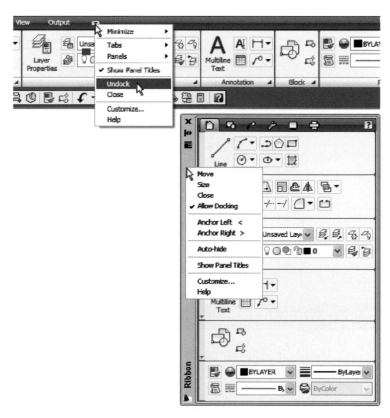

Fig. 1.33 The **Ribbon** as a palette with its *right-click* menu

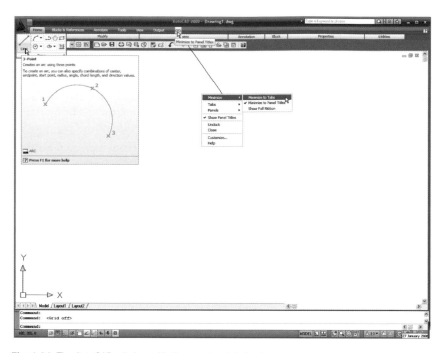

Fig. 1.34 The AutoCAD window with the panels minimized

The **Quick View Drawings** button

One of the buttons at the right-hand end of the status bar is the **Quick View Drawings** button. A *click* on this button brings miniatures of recent drawings on screen (Fig. 1.35). This can be of value when wishing to check back features of recent drawings in relation to the current drawing on screen.

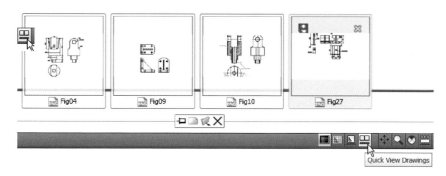

Fig. 1.35 The result of a *click* on the **Quick View Drawings** button

Customization of user interface

The AutoCAD 2009 workspace can be arranged in any format the operator wishes by making settings in the **Customize User Interface** dialog (Fig. 1.36) which can be brought on to screen from the *right-click* menu next to the toolbars at the top of the AutoCAD window. The dialog can be opened using other methods such as *entering* **cui** at the command line, but

Fig. 1.36 The **Customize User Interface** dialog brought to screen from the **right-click** menu next to the toolbars

using this *right-click* menu is possibly the quickest method. The dialog is only shown here to alert the reader to the fact that he/she can customize the workspace being used to suit their own methods of working. Space limitations in this book do not allow further explanation.

REVISION NOTES

1. A *double-click* on the **AutoCAD 2009** shortcut in the Windows desktop opens the AutoCAD window.
2. Alternatively, *right-click* on the shortcut, followed by a *left-click* on **Open** in the menu which then appears.
3. There are three main workspaces in which drawings can be constructed – the **2D Drafting & Annotation**, **Classic AutoCAD** and the **3D Modeling** workspace. Part 1 of this book deals with 2D drawings, which will be constructed mainly in the **2D Drafting & Annotation** workspace. In Part 2 of this book, 3D model drawings will be mainly constructed in the **3D Modeling** workspace.
4. All constructions in this book involve the use of a mouse as the digitizer. When a mouse is the digitizer:
    - a *left-click* means pressing the left-hand button (the **Pick** button)
    - a *right-click* means pressing the right-hand button (the **Return** button)

- a *double-click* means pressing the left-hand button twice in quick succession
- *dragging* means moving the mouse until the cursor is over an item on screen, holding the left-hand button down and moving the mouse. The item moves in sympathy to the mouse movement
- to *pick* has a similar meaning to *left-click*.

6. Palettes are a particular feature of AutoCAD 2009. The **Command** palette and the **DesignCenter** palette will be in frequent use.
7. Tools are shown as icons in the tool panels.
8. When a tool is *picked*, a tooltip describing the tool appears, which describes the action of the tool.
9. Dialogs allow opening and saving of files and the setting of parameters.
10. A number of *right-click* menus are used in AutoCAD 2009.
11. A number of buttons in the status bar can be used to toggle features such as snap and grid. Functions keys of the keyboard can be also used for toggling some of these functions.
12. The AutoCAD coordinate system determines the position in units of any 2D point in the drawing area (**2D Drafting & Annotation** and **Classic AutoCAD**) and any point in 3D space (**3D Modeling**).
13. Drawings are usually constructed in templates with predetermined settings. Some templates include borders and title blocks.

Note

Throughout this book when tools are to be selected from panels in the ribbon the tools will be shown in the following form: e.g. **Home/Draw** – the name of the tab in the ribbon title bar, followed by the name of the panel from which the tool is to be selected.

Chapter 2

Introducing drawing

AIMS OF THIS CHAPTER

The aims of this chapter are:

1. to introduce the construction of 2D drawing in the **2D Drafting & Annotation** workspace;
2. to introduce the drawing of outlines using the **Line, Circle** and **Polyline** tools from the **Home/Draw** panel;
3. to introduce drawing to snap points;
4. to introduce drawing to absolute coordinate points;
5. to introduce drawing to relative coordinate points;
6. to introduce drawing using the 'tracking' method;
7. to introduce the use of the **Erase, Undo** and **Redo** tools.

The 2D Drafting & Annotation workspace

Illustrations throughout this chapter will be shown using the **2D Drafting & Annotation** workspace. However, methods of construction will be the same if the reader wishes to work in other workspaces. In this workspace, the **Home/Draw** panel is at the left-hand end of the **Ribbon**, and **Draw** tools can be selected from the panel as indicated by a *click* on the **Line** tool (Fig. 2.1). In this chapter all examples will show tools as selected from the **Home/Draw** panel. It is possible to select tools from the **Draw** toolbar. To place the **Draw** toolbar on screen when working in the **2D Drafting & Annotation** workspace, *right-click* in the area to the right of the **Standard Annotation** toolbar, followed by a *right-click* on **AutoCAD** in the menu which appears, and selecting **Draw** from the list of toolbars. The **Line** tool with its tooltip is shown, selected from the **Draw** toolbar in Fig. 2.2.

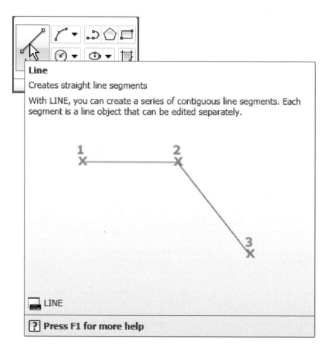

Fig. 2.1 The **Line** tool from the **Home/Draw** panel with its tooltip

Any toolbar can also be selected from the *right-click* menu in any toolbar on screen as shown in Fig. 2.2.

Drawing with the **Line** tool

First example – **Line** tool (Fig. 2.3)

1. Open AutoCAD. The drawing area will show the settings of the **acadiso.dwt** template – **Limits** set to **420,297**, **Grid** set to **10**, **Snap** set to **5** and **Units** set to **0**.

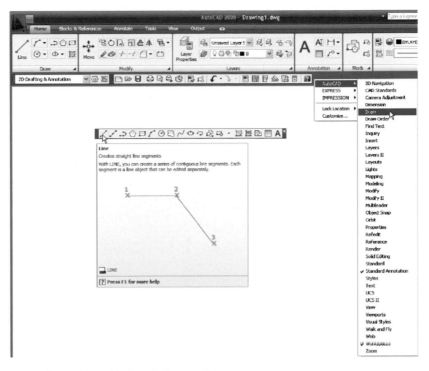

Fig. 2.2 **Line** and its tooltip from the **Draw** toolbar

2. *Left-click* on the **Line** tool in the **Home/Draw** panel or **Draw** toolbar (Figs 2.1 and 2.2).

Note

(a) The tooltip which appears when the tool icon is *clicked*.

(b) The prompt **Command:_line Specify first point** which appears in the command window at the command line (Fig. 2.3).

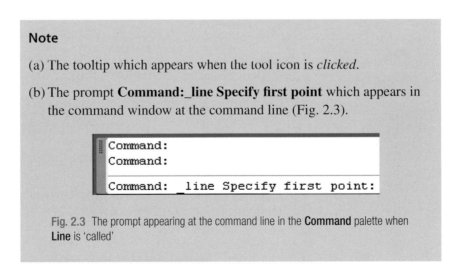

Fig. 2.3 The prompt appearing at the command line in the **Command** palette when **Line** is 'called'

3. Make sure **Snap** is on by either pressing the **F9** key or the **Snap Mode** button in the status bar. <**Snap on**> will show in the command palette.

4. Move the mouse around the drawing area. The cursors pick box will jump from point to point at 5 unit intervals. The position of the pick box will show as coordinate numbers in the status bar (left-hand end).

5. Move the mouse until the coordinate numbers show **60,240,0** and press the *pick* button of the mouse (*left-click*).
6. Move the mouse until the coordinate numbers show **260,240,0** and *left-click.*
7. Move the mouse until the coordinate numbers show **260,110,0** and *left-click.*
8. Move the mouse until the coordinate numbers show **60,110,0** and *left-click.*
9. Move the mouse until the coordinate numbers show **60,240,0** and *left-click.* Then press the **Return** button of the mouse (*right-click*). The line rectangle (Fig. 2.4) appears in the drawing area.

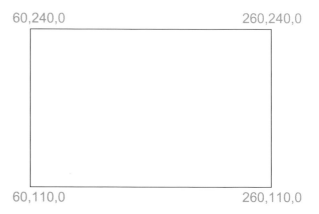

60,240,0   260,240,0

60,110,0   260,110,0

Fig. 2.4 First example – **Line** tool

Second example – **Line** tool (Fig. 2.6)

1. Clear the drawing from the screen with a *click* on the drawing **Close** button of the AutoCAD drawing area. Make sure it is not the AutoCAD 2009 window button.
2. The warning window (Fig. 2.5) appears in the centre of the screen. *Click* its **No** button.
3. *Left-click* on **New…** in the **File** drop-down menu and from the **Select template** dialog which appears *double-click* on **acadiso.dwt**.

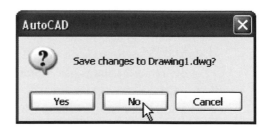

Fig. 2.5 The **AutoCAD** warning window

4. *Left-click* on the **Line** tool icon and *enter* figures as follows at each prompt of the command line sequence:

```
Command: _line Specify first point: enter 80,235
  right-click
Specify next point or [Undo]: enter 275,235
  right-click
Specify next point or [Undo]: enter 295,210
  right-click
Specify next point or [Close/Undo]: enter
  295,100 right-click
Specify next point or [Close/Undo]: enter
  230,100 right-click
Specify next point or [Close/Undo]: enter
  230,70 right-click
Specify next point or [Close/Undo]: enter
  120,70 right-click
Specify next point or [Close/Undo]: enter
  120,100 right-click
Specify next point or [Close/Undo]: enter
  55,100 right-click
Specify next point or [Close/Undo]: enter
  55,210 right-click
Specify next point or [Close/Undo]: enter c
  (Close) right-click
Command:
```

The result is as shown in Fig. 2.6.

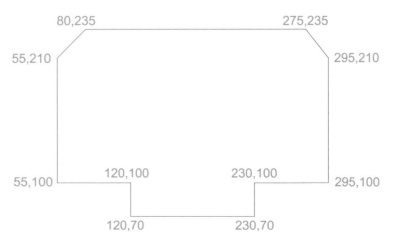

Fig. 2.6 Second example – **Line** tool

Third example – **Line** tool (Fig. 2.7)

1. Close the drawing and open a new **acadiso.dwt** window.
2. *Left-click* on the **Line** tool icon and *enter* figures as follows at each prompt of the command line sequence:

```
Command: _line Specify first point: enter 60,210
   right-click
Specify next point or [Undo]: enter @50,0
   right-click
Specify next point or [Undo]: enter @0,20
   right-click
Specify next point or [Close/Undo]: enter @130,0
   right-click
Specify next point or [Close/Undo]: enter
   @0,-20 right-click
Specify next point or [Close/Undo]: enter @50,0
   right-click
Specify next point or [Close/Undo]: enter
   @0,-105 right-click
Specify next point or [Close/Undo]: enter
   @-50,0 right-click
Specify next point or [Close/Undo]: enter
   @0,-20 right-click
Specify next point or [Close/Undo]: enter
   @-130,0 right-click
Specify next point or [Close/Undo]: enter @0,20
   right-click
Specify next point or [Close/Undo]: enter
   @-50,0 right-click
Specify next point or [Close/Undo]: enter c
   (Close) right-click
Command:
```

The result is as shown in Fig. 2.7.

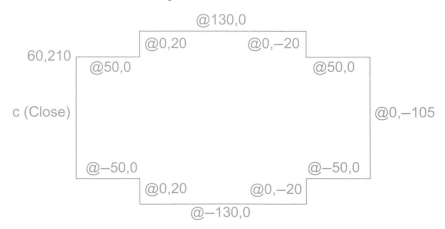

Fig. 2.7 Third example –**Line** tool

Notes

1. The figures typed at the keyboard determining the corners of the outlines in the above examples are two-dimensional (2D) **x,y** coordinate points. When working in 2D, coordinates are expressed in terms of two numbers separated by a comma.

2. Coordinate points can be shown as positive or as negative numbers.

3. The method of constructing an outline as shown in the first two examples above is known as the **absolute coordinate entry** method, where the **x,y** coordinates of each corner of the outlines are *entered* at the command line as required.

4. The method of constructing an outline as in the third example is known as the **relative coordinate entry** method – coordinate points are *entered* relative to the previous entry. In relative coordinate entry, the @ symbol is *entered* before each set of coordinates with the following rules in mind:
   +ve x entry is to the right
   −ve x entry is to the left
   +ve y entry is upwards
   −ve y entry is downwards.

5. The next example (the fourth) shows how lines at angles can be drawn taking advantage of the relative coordinate entry method. Angles in AutoCAD are measured in 360 degrees in a counter-clockwise (anticlockwise) direction (Fig. 2.8). The < symbol precedes the angle.

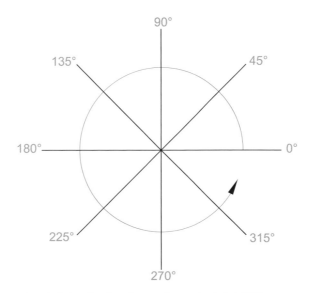

Fig. 2.8 The counter-clockwise direction of measuring angles in AutoCAD

Fourth example – **Line** tool (Fig. 2.9)

1. Close the drawing and open a new **acadiso.dwt** window.
2. *Left-click* on the **Line** tool icon and *enter* figures as follows at each prompt of the command line sequence:

```
Command: _line Specify first point: 70,230
Specify next point: @220,0
Specify next point: @0,-70
Specify next point or [Undo]: @115<225
Specify next point or [Undo]: @-60,0
Specify next point or [Close/Undo]: @115<135
Specify next point or [Close/Undo]: @0,70
Specify next point or [Close/Undo]: c (Close)
Command:
```

The result is as shown in Fig. 2.9.

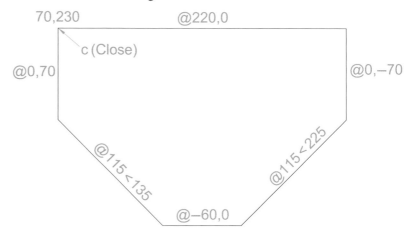

Fig. 2.9 Fourth example – **Line** tool

Fifth example – **Line** tool (Fig. 2.10)

Another method of constructing accurate drawings is by using a method known as **tracking.** When **Line** is in use, as each **Specify next point:** appears at the command line, a *rubber-banded* line appears from the last point *entered*. *Drag* the rubber-band line in any direction and *enter* a number at the keyboard, followed by a *right-click*. The line is drawn in the *dragged* direction of a length in units equal to the *entered* number.

In this example, because all lines are drawn in vertical or horizontal directions, either press the **F8** key or *click* the **ORTHO** button in the status bar which will only allow drawing horizontally or vertically:

1. Close the drawing and open a new **acadiso.dwt** window.
2. *Left-click* on the **Line** tool icon and *enter* figures as follows at each prompt of the command line sequence:

```
Command: _line Specify first point: enter 65,220
   right-click
```

```
Specify next point: drag to right enter 240
  right-click
Specify next point: drag down enter 145
  right-click
Specify next point or [Undo]: drag left enter
  65 right-click
Specify next point or [Undo]: drag upwards
  enter 25 right-click
Specify next point or [Close/Undo]: drag left
  enter 120 right-click
Specify next point or [Close/Undo]: drag upwards
  enter 25 right-click
Specify next point or [Close/Undo]: drag left
  enter 55 right-click
Specify next point or [Close/Undo]: c (Close)
  right-click
Command:
```

The result is as shown in Fig. 2.10.

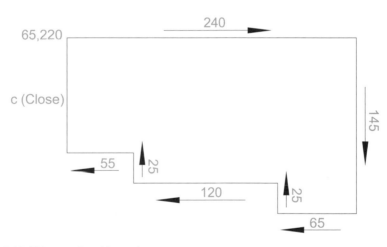

Fig. 2.10 Fifth example – **Line** tool

Drawing with the Circle tool

First example – **Circle** tool (Fig. 2.13)

1. Close the drawing just completed and open the **acadiso.dwt** template.
2. *Left-click* on the **Circle** tool icon in the **Home/Draw** panel (Fig. 2.11).
3. *Enter* a coordinate and a radius against the prompts appearing in the command window as shown in Fig. 2.12, followed by *right-clicks*. The circle (Fig. 2.13) appears on screen.

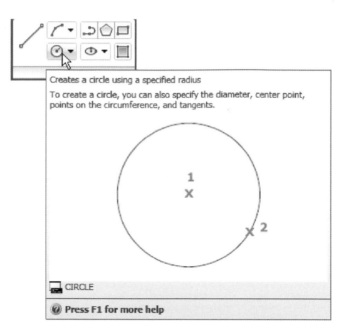

Fig. 2.11 The **Circle** tool from the **Home/Draw** panel

```
Command: _circle Specify center point for circle or [3P/2P/Ttr (tan tan
radius)]: 180,160
Specify radius of circle or [Diameter] <40.0000>: 55
Command:
```

Fig. 2.12 First example – **Circle**. The command line prompts when **Circle** is called

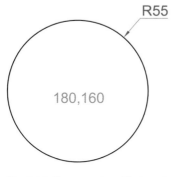

Fig. 2.13 First example – **Circle** tool

Second example – **Circle** tool (Fig. 2.14)

1. Close the drawing and open the **acadiso.dwt** screen.
2. *Left-click* on the **Circle** tool icon and construct two circles as shown in the drawing (Fig. 2.14) in the positions and radii shown in Fig. 2.15.
3. *Click* the **Circle** tool again and against the first prompt *enter* **t** (the abbreviation for the prompt **tan tan radius**), followed by a *right-click*.

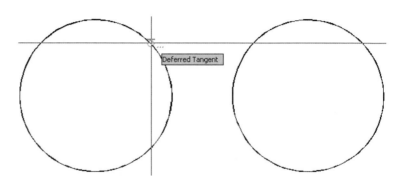

Fig. 2.14 Second example – **Circle** tool – the two circles of radius 50

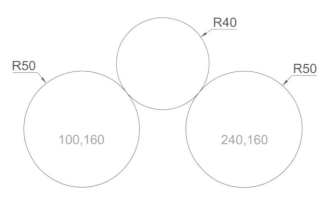

Fig. 2.15 Second example – **Circle** tool. The radius-40 circle tangential to the radius-50 circles

```
Command_circle Specify center point for circle or
  [3P/2P/Ttr (tan tan radius)]: enter t right-click
Specify point on object for first tangent of
  circle: pick
Specify point on object for second tangent of
  circle: pick
Specify radius of circle (50): enter 40
  right-click
Command:
```

The radius-40 circle tangential to the two circles already drawn then appears (Fig. 2.15).

Notes

1. When a point on either circle is picked, a tip appears **Deferred Tangent**. This tip will only appear when the **Object Snap** button is set on with a *click* on its button in the status bar, or the **F3** key of the keyboard is pressed.

2. Circles can be drawn through 3 points or through 2 points *entered* at the command line in response to prompts brought to the command line by using **3P** and **2P** in answer to the circle command line prompts.

The Erase tool

If an error has been made when using any of the AutoCAD 2009 tools, the object or objects which have been incorrectly drawn can be deleted with the **Erase** tool. The **Erase** tool icon can be selected from the **Home/Modify** panel (Fig. 2.16) or by *entering* **e** at the command line.

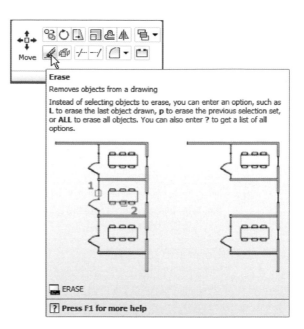

Fig. 2.16 The **Erase** tool icon from the **Home/Modify** panel

First example – Erase (Fig. 2.18)

1. With **Line** construct the outline (Fig. 2.17).

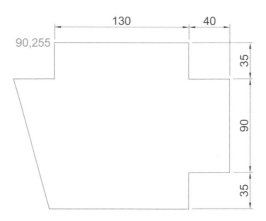

Fig. 2.17 First example – **Erase**. An incorrect outline

2. Assuming two lines of the outline have been incorrectly drawn, *left-click* the **Erase** tool icon. The command line shows:

```
Command: _erase
Select objects: pick one of the lines 1 found
Select objects: pick the other line 2 total
Select objects: right-click
Command:
```

and the two lines are deleted (right-hand drawing of Fig. 2.18).

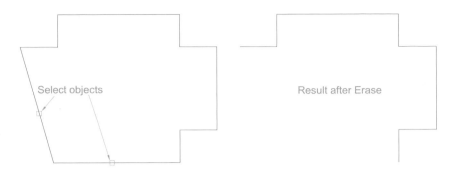

Fig. 2.18 First example – **Erase**

Second example – **Erase** (Fig. 2.19)

The two lines could also have been deleted by the following method:

1. *Left-click* the **Erase** tool icon. The command line shows:

```
Command: _erase
Select objects: enter c (Crossing)
Specify first corner: pick Specify opposite
  corner: pick 2 found
Select objects: right-click
Command:
```

and the two lines are deleted as in the right-hand drawing of Fig. 2.19.

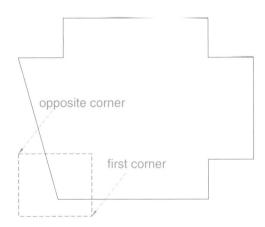

Fig. 2.19 Second example – **Erase**

Undo and Redo tools

Two other tools of value when errors have been made are the **Undo** and **Redo** tools. To undo the last action taken by any tool when constructing a drawing, either *left-click* the **Undo** tool in the **Standard Annotation** toolbar (Fig. 2.20) or *enter* **u** at the command line. Regardless of which method is adopted the error is deleted from the drawing.

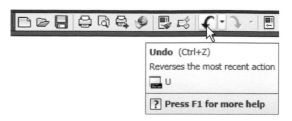

Fig. 2.20 The **Undo** tool in the **Standard Annotation** toolbar

Everything done during a session in constructing a drawing can be undone by repeated *clicking* on the **Undo** tool icon or by *entering* **u**'s at the command line.

To bring back objects that have just been removed by the use of **Undo**'s, *left-click* the **Redo** tool icon in the **Standard Annotation** toolbar (Fig. 2.21) or *enter* **Redo** at the command line.

Drawing with the **Polyline** tool

Fig. 2.21 The **Redo** tool icon in the **Standard Annotation** toolbar

When drawing lines with the **Line** tool, each line drawn is an object in its own right. A rectangle drawn with the **Line** tool is four objects. A rectangle drawn with the **Polyline** tool is a single object. Lines of different thickness, arcs, arrows and circles can all be drawn using this tool as will be shown in the examples describing constructions using the **Polyline** tool. Constructions resulting from using the tool are known as **polylines** or **plines**.

The **Polyline** tool can be called from the **Home/Draw** panel (Fig. 2.22) or by *entering* **pl** at the command line.

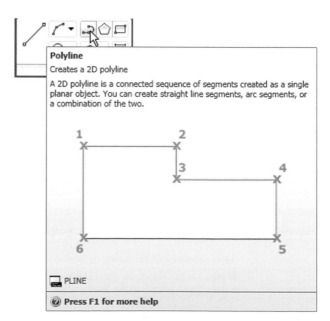

Fig. 2.22 The **Polyline** tool icon in the **Home/Draw** panel

First example – **Polyline** tool (Fig. 2.23)

Fig. 2.23 First example – **Polyline** tool

Note

In this example *enter* and *right-click* have not been included.

Left-click the **Polyline** tool. The command line shows:

```
Command: _pline Specify start point: 30,250
Current line width is 0
Specify next point or [Arc/Halfwidth/Length/
Undo/Width]: 230,250
Specify next point or [Arc/Close/Halfwidth/
Length/Undo/Width]: 230,120
Specify next point or [Arc/Close/Halfwidth/
Length/Undo/Width]: 30,120
Specify next point or [Arc/Close/Halfwidth/
Length/Undo/Width]: c (Close)
Command:
```

Note

1. Note the prompts – **Arc** for constructing pline arcs; **Close** to close an outline; **Halfwidth** to halve the width of a wide pline; **Length** to *enter* the required length of a pline; **Undo** to undo the last pline constructed; **Close** to close an outline.

2. Only the capital letter(s) of a prompt needs to be *entered* in upper or lower case to make that prompt effective.

3. Other prompts will appear when the **Polyline** tool is in use as will be shown in later examples.

Second example – **Polyline** tool (Fig. 2.24)

This will be a long sequence, but it is typical of a reasonably complex drawing using the **Polyline** tool. In the following sequences, when a prompt line is to be repeated, the prompts in square brackets ([]) will be replaced by **[prompts]**.

Left-click the **Polyline** tool icon. The command line shows:

```
Command: _pline Specify start point: 40,250
Current line width is 0
Specify next point or [Arc/Halfwidth/Length/Undo/
  Width]: w (Width)
Specify starting width <0>: 5
Specify ending width <5>: right-click
Specify next point or [Arc/Close/Halfwidth/Length/
  Undo/Width]: 160,250
Specify next point or [prompts]: h (Halfwidth)
Specify starting half-width <2.5>: 1
Specify ending half-width <1>: right-click
Specify next point or [prompts]: 260,250
Specify next point or [prompts]: 260,180
Specify next point or [prompts]: w (Width)
Specify starting width <1>: 10
Specify ending width <10>: right-click
Specify next point or [prompts]: 260,120
Specify next point or [prompts]: h (Halfwidth)
Specify starting half-width <5>: 2
Specify ending half-width <2>: right-click
Specify next point or [prompts]: 160,120
Specify next point or [prompts]: w (Width)
Specify starting width <4>: 20
Specify ending width <20>: right-click
Specify next point or [prompts]: 40,120
Specify starting width <20>: 5
Specify ending width <5>: right-click
Specify next point or [prompts]: c (Close)
Command:
```

Fig. 2.24 Second example – **Polyline** tool

Third example – **Polyline** tool (Fig. 2.25)

Left-click the **Polyline** tool icon. The command line shows:

```
Command: _pline Specify start point: 50,220
Current line width is 0
  [prompts]: w (Width)
Specify starting width <0>: 0.5
Specify ending width <0.5>: right-click
Specify next point or [prompts]: 120,220
Specify next point or [prompts]: a (Arc)
Specify endpoint of arc or [prompts]: s (second pt)
Specify second point on arc: 150,200
Specify end point of arc: 180,220
Specify end point of arc or [prompts]: l (Line)
Specify next point or [prompts]: 250,220
Specify next point or [prompts]: 250,190
Specify next point or [prompts]: a (Arc)
Specify endpoint of arc or [prompts]: s (second pt)
Specify second point on arc: 240,170
Specify end point of arc: 250,150
Specify end point of arc or [prompts]: l (Line)
Specify next point or [prompts]: 250,150
Specify next point or [prompts]: 250,120
Command:
```

and so on until the outline in Fig. 2.25 is completed.

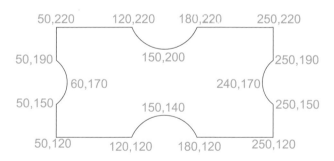

Fig. 2.25 Third example – **Polyline** tool

Fourth example – **Polyline** tool (Fig. 2.26)

Left-click the **Polyline** tool icon. The command line shows:

```
Command: _pline Specify start point: 80,170
Current line width is 0
Specify next point or [prompts]: w (Width)
```

```
Specify starting width <0>: 1
Specify ending width <1>: right-click
Specify next point or [prompts]: a (Arc)
Specify endpoint of arc or [prompts]: s (second pt)
Specify second point on arc: 160,250
Specify end point of arc: 240,170
Specify end point of arc or [prompts]: cl (Close)
Command:
```

and the circle in Fig. 2.26 is formed.

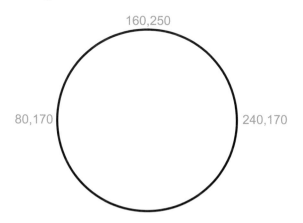

Fig. 2.26 Fourth example – **Polyline** tool

Fifth example – **Polyline** tool (Fig. 2.27)

Left-click the **Polyline** tool icon. The command line shows:

```
Command: _pline Specify start point: 60,180
Current line width is 0
Specify next point or [prompts]: w (Width)
Specify starting width <0>: 1
Specify ending width <1>: right-click
Specify next point or [prompts]: 190,180
Specify next point or [prompts]: w (Width)
Specify starting width <1>: 20
Specify ending width <20>: 0
Specify next point or [prompts]: 265,180
Specify next point or [prompts]: right-click
Command:
```

and the arrow in Fig. 2.27 is formed.

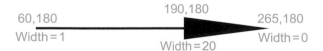

Fig. 2.27 Fifth example – **Polyline** tool

REVISION NOTES

1. The following terms have been used in this chapter:

- **_Left-click_** – press the left-hand button of the mouse.
- **_Click_** – same meaning as _left-click._
- **_Double-click_** – press the left-hand button of the mouse twice.
- **_Right-click_** – press the left-hand button of the mouse – has the same result as pressing the **Return** key of the keyboard.
- **_Drag_** – move the cursor on to an object and, holding down the right-hand button of the mouse, pull the object to a new position.
- **_Enter_** – type the letters or numbers which follow at the keyboard.
- **_Pick_** – move the cursor on to an item on screen and press the left-hand button of the mouse.
- **Return** – press the **Enter** key of the keyboard. This key may also be marked with a left-facing arrow. In most cases (but not always) it has the same result as a _right-click._
- **Dialog** – a window appearing in the AutoCAD window in which settings may be made.
- **Drop-down menu** – a menu appearing when one of the names in the menu bars is _clicked._
- **Tooltip** – the name of a tool appearing when the cursor is placed over a tool icon from a toolbar.
- **Prompts** – text appearing in the command window when a tool is selected which advises the operator as to which operation is required.

2. Three methods of coordinate entry have been used in this chapter:

- **Absolute method** – the coordinates of points on an outline are _entered_ at the command line in response to prompts.
- **Relative method** – the distances in coordinate units are _entered_ preceded by **@** from the last point which has been determined on an outline. Angles, which are measured in a counter-clockwise direction, are preceded by a point (.).
- **Tracking** – the rubber band of the tool is _dragged_ in the direction in which the line is to be drawn and its distance in units is _entered_ at the command line followed by a _right-click._
- **Line and Polyline tools** – an outline drawn using the **Line** tool consists of a number of objects. the number of lines in the outline. An outline drawn using the **Polyline** is a single object.

Exercises

Methods of constructing answers to the following exercises can be found in the free website:

http://books.elsevier. com/companions/9780750689830

1. Using the **Line** tool construct the rectangle shown in Fig. 2.28.

Fig. 2.28 Exercise 1

2. Construct the outline shown in Fig. 2.29 using the **Line** tool. The coordinate points of each corner of the rectangle will need to be calculated from the lengths of the lines between the corners.

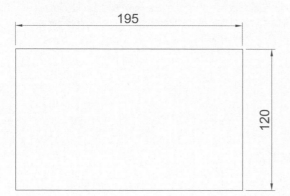

Fig. 2.29 Exercise 2

3. Using the **Line** tool, construct the outline shown in Fig. 2.30.

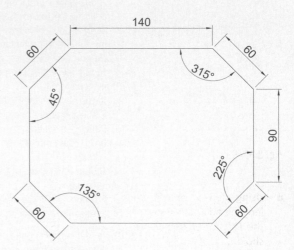

Fig. 2.30 Exercise 3

4. Using the **Circle** tool, construct the two circles of radius 50 and 30. Then using the **Ttr** prompt add the circle of radius 25 and then add the circle of radius 25 (Fig. 2.31).

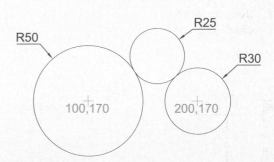

Fig. 2.31 Exercise 4

5. In an **acadiso.dwt** screen and using the **Circle** and **Line** tools, construct the line and the circle of radius 40 shown in Fig. 2.32. Then, using the **Ttr** prompt, add the circle of radius 25.

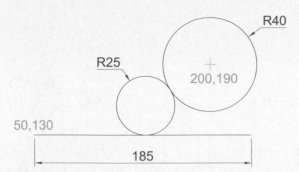

Fig. 2.32 Exercise 5

6. Using the **Line** tool construct the two lines at the length and angle as given in Fig. 2.33. Then, with the **Ttr** prompt of the **Circle** tool, add the circle as shown.

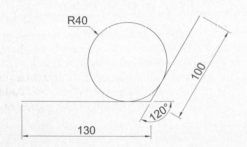

Fig. 2.33 Exercise 6

7. Using the **Polyline** tool, construct the outline given in Fig. 2.34.

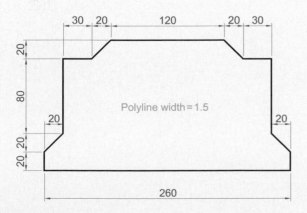

Fig. 2.34 Exercise 7

8. Construct the outline given in Fig. 2.35 using the **Polyline** tool.

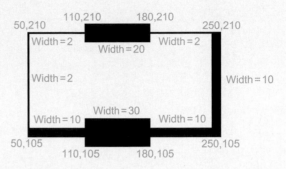

Fig. 2.35 Exercise 8

9. With the **Polyline** tool construct the arrows shown in Fig. 2.36.

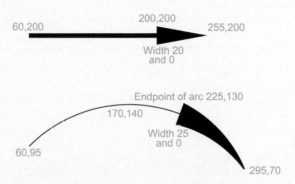

Fig. 2.36 Exercise 9

Chapter 3

Draw tools, Object Snap and Dynamic Input

AIMS OF THIS CHAPTER

The aims of this chapter are:

1. to give examples of the use of the **Arc, Ellipse**, **Polygon** and **Rectangle** tools from the **Home/Draw** panel;
2. to give examples of the uses of the **Polyline Edit (pedit)** tool;
3. to introduce the **Object Snaps (Osnaps)** and their uses;
4. to introduce the **Dynamic Input (DYN)** system and its uses.

Introduction

The majority of tools in AutoCAD 2009 can be called into use by any one of the following five methods:

1. placing the cursor on the tool's icon in the appropriate panel. Figure 3.2 shows the **Polygon** tool selected from the **Home/Draw** panel;
2. with a *click* on the tool's name in a toolbar. Figure 3.3 shows the **Draw** toolbar. Placing the cursor on the **Polygon** tool icon in this toolbar shows the same tooltip as that shown in Fig. 3.2;
3. by *clicking* on the tool's name in an appropriate drop-down menu. Figure 3.4 shows the tool names and icons displayed in the **Draw** drop-down menu. It is necessary to first bring the menu bar to screen with a *click* on **Show Menu Bar** in the *right-click* menu of the **Quick Access Toolbar** (Fig. 3.1);

Fig. 3.1 Bringing the menu bar on screen

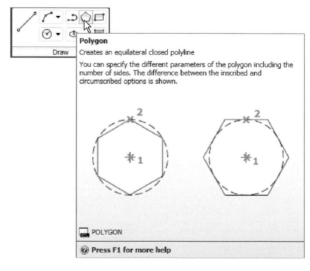

Fig. 3.2 The **Polygon** tool and its tooltip selected from the **Home/Draw** panel

Fig. 3.3 The tool icons in the **Draw** toolbar

4. by *entering* an abbreviation for the tool name at the command line in the command palette. For example the abbreviation for the **Line** tool is **l**, for the **Polyline** tool it is **pl** and for the **Circle** tool it is **c**;
5. by *entering* the full name of the tool at the command line.

Fig. 3.4 The **Draw** drop-down menu

In practice, operators constructing drawings in AutoCAD 2009 may well use a combination of these five methods.

The **Arc** tool

In AutoCAD 2009, arcs can be constructed using any three of the following characteristics of an arc – its **Start** point; a point on the arc (**Second** point); its **Center**; its **End**; its **Radius**; the **Length** of the arc; the **Direction** in which the arc is to be constructed; the **Angle** between lines of the arc.

These characteristics are shown in the menu appearing with a *click* on the **Arc** tool in the **Home/Draw** panel (Fig. 3.5).

To call the **Arc** tool *click* on its tool icon in the **Home/Draw** panel, *click* on **Arc** in the **Draw** toolbar, *click* on **Arc** in the **Draw** drop-down menu, or *enter* **a** or **arc** at the command line. In the following examples initials of command prompts will be shown instead of selection from the menu shown in Fig. 3.6.

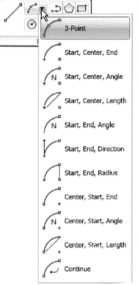

Fig. 3.6 The menu appearing with a *click* on the arrow next to the **Arc** tool icon in the **Home/Draw** panel

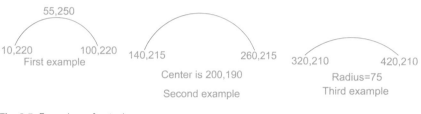

Fig. 3.5 Examples – **Arc** tool

First example – **Arc** tool (Fig. 3.5)

Left-click the **Arc** tool icon. The command line shows:

```
Command: _arc Specify start point of arc or
  [Center]: 100,220
Specify second point of arc or [Center/End]: 55,250
Specify end point of arc: 10,220
Command:
```

Second example – **Arc** tool (Fig. 3.5)

```
Command: right-click brings back the Arc sequence
ARC Specify start point of arc or [Center]:
  c (Center)
Specify center point of arc: 200,190
Specify start point of arc: 260,215
Specify end point of arc or [Angle/chord Length]:
  140,215
Command:
```

Third example – **Arc** tool (Fig. 3.5)

```
Command: right-click brings back the Arc sequence
ARC Specify start point of arc or [Center]: 420,210
Specify second point of arc or [Center/End]: e (End)
Specify end point of arc: 320,210
Specify center point of arc or [Angle/Direction/
  Radius]: r (Radius)
Specify radius of arc: 75
Command:
```

The **Ellipse** tool

Ellipses can be regarded as what is seen when a circle is viewed from directly in front of the circle and the circle rotated through an angle about its horizontal diameter. Ellipses are measured in terms of two axes – a **major axis** and a **minor axis**, the major axis being the diameter of the circle, the minor axis being the height of the ellipse after the circle has been rotated through an angle (Fig. 3.7).

To call the **Ellipse** tool, *click* on its tool icon in the **Home/Draw** panel (Fig. 3.8), *click* its name in the **Draw** drop-down menu, *click* on its tool icon in the **Draw** toolbar or *enter* **a** or **arc** at the command line.

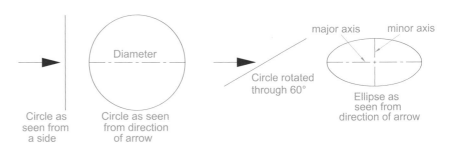

Fig. 3.7 An ellipse can be regarded as viewing a rotated circle

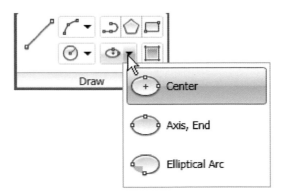

Fig. 3.8 The **Ellipse** tool icon flyout in the the **Home/Draw** panel

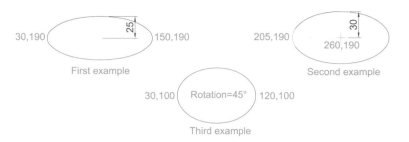

Fig. 3.9 Examples – **Ellipse**

First example – **Ellipse** (Fig. 3.9)

Left-click the **Ellipse** tool icon. The command line shows:

```
Command: _ellipse
Specify axis endpoint of elliptical arc or [Arc/
  Center]: 30,190
Specify other endpoint of axis: 150,190
Specify distance to other axis or [Rotation] 25
Command:
```

Second example – **Ellipse** (Fig. 3.9)

In this second example, the coordinates of the centre of the ellipse (the point where the two axes intersect) are *entered*, followed by *entering* coordinates for the end of the major axis, followed by *entering* the units for the end of the minor axis.

```
Command: right-click
ELLIPSE
Specify axis endpoint of elliptical arc or [Arc/
  Center]: c
Specify center of ellipse: 260,190
Specify endpoint of axis: 205,190
Specify distance to other axis or [Rotation]: 30
Command:
```

Third example – **Ellipse** (Fig. 3.9)

In this third example, after setting the positions of the ends of the major axis, the angle of rotation of the circle from which an ellipse can be obtained is *entered*.

```
Command: right-click
ELLIPSE
Specify axis endpoint of elliptical arc or [Arc/
  Center]: 30,100
Specify other endpoint of axis: 120,100
Specify distance to other axis or [Rotation]:
  r (Rotation)
Specify rotation around major axis: 45
Command:
```

Saving drawings

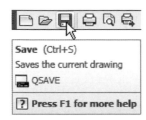

Fig. 3.10 Selecting **Save** in the **Standard Annotation** toolbar

Before going further it is as well to know how to save the drawings constructed when answering examples and exercises in this book. When a drawing has been constructed, *left-click* on the **Save** icon **Standard Annotation** toolbar (Fig. 3.10). The **Save Drawing As** dialog appears (Fig. 3.11).

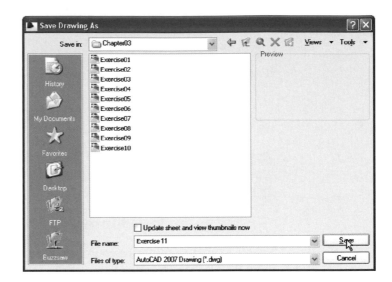

Fig. 3.11 The **Save Drawing As** dialog

Unless you are the only person to use the computer on which the drawing has been constructed, it is best to save work to a floppy disk, usually held in the drive **A:**. To save a drawing to a floppy in drive **A:**

1. Place a floppy disk in drive **A:**.
2. In the **Save in:** field of the dialog, *click* the arrow to the right of the field and from the pop-up list select **3½ Floppy [A:]**.
3. In the **File name:** field of the type a name. The file name extension **.dwg** does not need to be typed – it will be added to the file name.
4. *Left-click* the **Save** button of the dialog. The drawing will be saved with the file name extension **.dwg** – the AutoCAD file name extension.

> **Note**
>
> A **USB Flash drive** (pen) can also be used for this purpose.

Snap

In previous chapters several methods of constructing accurate drawings have been described – using **Snap**; absolute coordinate entry; relative coordinate entry; and tracking. Other methods of ensuring accuracy between parts of constructions are achieved by making use of **Object Snaps** (**Osnaps**).

Snap Mode, **Grid Display** and **Object Snaps** can be toggled on/off from the buttons in the status bar or by pressing the keys, **F9** (**Snap Mode**), **F7** (**Grid Display**) and **F3** (**Object Snap**).

Object Snaps (Osnaps)

Object Snaps allow objects to be added to a drawing at precise positions in relation to other objects already on screen. With **Object Snaps**, objects can be added to the endpoints and midpoints, to intersections of objects, to centres and quadrants of circles and so on. **Object Snaps** also override snap points even when snap is set on.

To set **Object Snaps** – at the command line:

Command: *enter os*

and the **Drafting Settings** dialog appears (Fig. 3.12). *Click* the **Object Snap** tab in the upper part of the dialog and *click* the check boxes to the right of the **Object Snap** names to set them on (or off in on).

Fig. 3.12 The **Drafting Settings** dialog with most of the **Object Snaps** set on

When **Object Snaps** are set **ON**, as outlines are constructed using **Object Snaps** so **Object Snap** icons and their tooltips appear as indicated in Fig. 3.13.

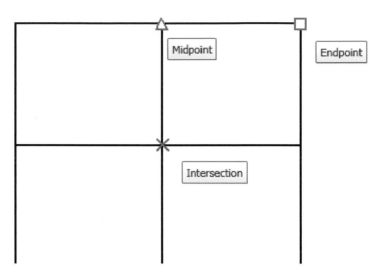

Fig. 3.13 Three **Object Snap** icons and their tooltips

It is sometimes advisable not to have **Object Snaps** set on in the **Drafting Settings** dialog, but to set **Object Snap** off and use **Object Snap** abbreviations at the command line when using tools. The following examples show the use of some of these abbreviations.

First example – **Object Snap** abbreviations (Fig. 3.14)

Call the **Polyline** tool:

```
Command: _pline
Specify start point: 50,230
[prompts]: w (Width)
Specify starting width: 1
Specify ending width <1>: right-click
Specify next point: 260,230
Specify next point: right-click
Command: right-click
PLINE
Specify start point: end of pick the right-hand
  end of the pline
Specify next point: 50,120
Specify next point: right-click
Command: right-click
PLINE
Specify start point: mid of pick near the middle
  of first pline
Specify next point: 155,120
Specify next point: right-click
Command: right-click
```

```
PLINE
Specify start point: int of pick the plines at
  their intersection
Specify start point: right-click
Command:
```

The result is shown in Fig. 3.14. In this illustration the **Object Snap** tooltips are shown as they appear when each object is added to the outline.

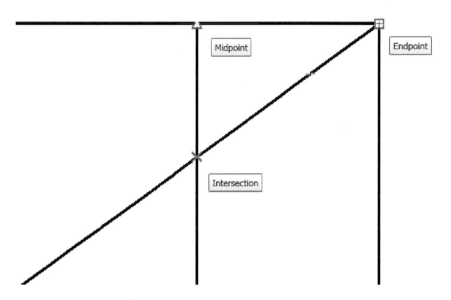

Fig. 3.14 First example – **Osnaps**

Second example – **Object Snap** abbreviations (Fig. 3.15)

Call the **Circle** tool:

```
Command: _circle
Specify center point for circle: 180,170
Specify radius of circle: 60
Command: enter l (Line) right-click
Specify first point: enter qua right-click
of pick near the upper quadrant of the circle
Specify next point: enter cen right-click
of pick near the center of the circle
Specify next point: enter qua right-click
of pick near right-hand side of circle
Specify next point: right-click
Command:
```

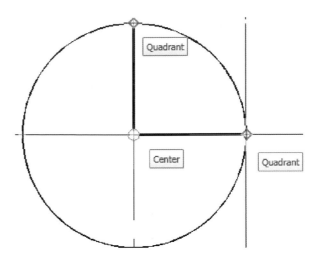

Fig. 3.15 Second example – **Osnaps**

Note

With **Object Snaps** off, the following abbreviations can be used:

end – endpoint
mid – midpoint
int – intersection
cen – center
qua – quadrant
nea – nearest
ext – extension.

Dynamic Input (DYN)

When **Dynamic Input** is set on by either pressing the **F12** key or with a *click* on the **Dynamic Input** button in the status bar dimensions, coordinate positions and commands appear as tips when no tool is in action (Fig. 3.16).

Fig. 3.16 The **DYN** tips appearing when no tool is in action and the cursor is moved

With a tool in action, as the cursor hairs are moved in response to movement of the mouse, **Dynamic Input** tips showing the coordinate figures for the point of the cursor hairs will show (Fig. 3.17), together with other details. To see the drop-down menu giving the prompts available with **Dynamic Input** press the **down** key of the keyboard and *click* the prompt to be used. Fig. 3.17 shows the **Arc** prompt as being the next to be used.

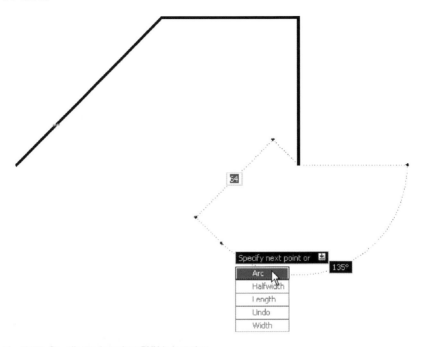

Fig. 3.17   Coordinate tips when **DYN** is in action

Notes on the use of **Dynamic Input**

Although **Dynamic Input** can be used in any of the three AutoCAD 2009 workspaces, some operators may prefer a larger working area. To achieve this a *click* on the **Clean Screen** icon in the bottom-right-hand corner of the AutoCAD 2009 window, produces an uncluttered workspace area. The command palette can be cleared from screen by *entering* **commandlinehide** at the command line. To bring it back press the keys **Ctrl+9**. These two operations produce a screen showing only title and status bars (Fig. 3.18). Some operators may well prefer working in such a larger than normal workspace.

Dynamic Input settings are made in the **Drafting Settings** dialog (Fig. 3.19), brought to screen by *entering* **ds** at the command line.

When **Dynamic Input** is in action, tools can be called by using any of the following methods:

1. by *entering* the name of the tool at the command line;
2. by *entering* the abbreviation for a tool name at the keyboard;

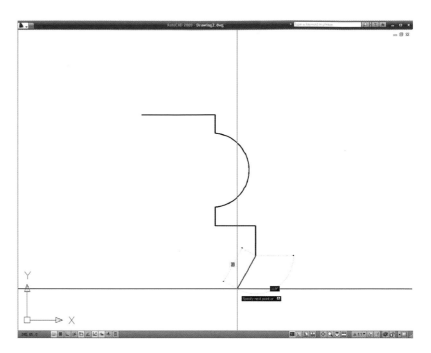

Fig. 3.18 Example of using **DYN** in a clear screen

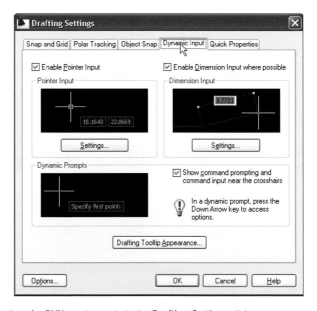

Fig. 3.19 Settings for **DYN** can be made in the **Drafting Settings** dialog

3. by selecting the tool's icon from a toolbar;
4. by selecting the tool's icon from a panel;
5. by selecting the tool's name from a drop-down menu.

When **Dynamic Input** is active and a tool is called, command prompts appear in a tooltip at the cursor position. Figure 3.20 shows the tooltip appearing at the cursor position when the **Line** tool icon in the **Home/Draw** panel is *clicked*.

Fig. 3.20 The prompt appearing on screen when the **Line** tool is selected

To commence drawing a line, either move the cursor under mouse control to the desired coordinate point and *left-click* as in Fig. 3.21, or *enter* the required *x,y* coordinates at the keyboard (Fig. 3.22) and *left-click*. To continue drawing with **Line** *drag* the cursor to a new position and either *left-click* at the position when the coordinates appear as required (Fig. 3.22), or *enter* a required length at the keyboard, which appears in the length box followed by a *left-click* (Fig. 3.23).

Fig. 3.21 *Drag* the cursor to the required point and *left-click*

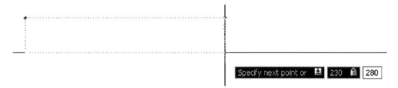

Fig. 3.22 *Enter* coordinates for the next point and *left-click*

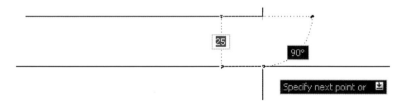

Fig. 3.23 *Enter* length at keyboard and *right-click*

Examples of using **Dynamic Input**

When using **Dynamic Input** the selection of a prompt can be made by pressing the **down** key of the keyboard (Fig. 3.24), which causes a pop-up menu to appear. A *click* on the required prompt in such a pop-up menu will make that prompt active.

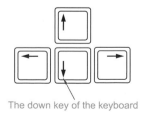

The down key of the keyboard

Fig. 3.24 The **down** key of the keyboard

Dynamic Input – first example – **Polyline**

1. Select **Polyline** from the **Home/Draw** panel (Fig. 3.25).
2. To start the construction *click* at any point on screen. The prompt for the **Polyline** appears with the coordinates of the selected point showing. *Left-click* to start the drawing (Fig. 3.26).

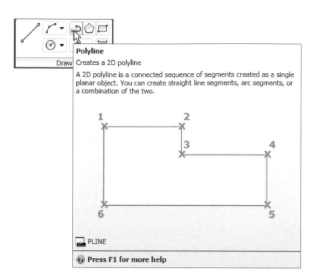

Fig. 3.25 **Dynamic Input** – first example – **Polyline** – selecting **Polyline** from the **Home**/**Draw** panel

Fig. 3.26 **Dynamic Input** – first example – **Polyline** – the first prompt

3. Move the cursor and press the **down** key of the keyboard. A pop-up menu appears from which a prompt selection can be made. In the menu *click* **Width** (Fig. 3.27).

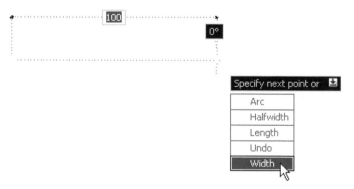

Fig. 3.27 **Dynamic Input** – first example – **Polyline** – *click* **Width** in the pop-up menu

4. Another prompt field appears. At the keyboard *enter* the required width and *right-click*. Then *left-click* and *enter* **ending width** or *right-click* if the **ending width** is the same as the **starting width** (Fig. 3.28).

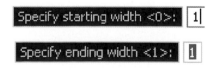

Fig. 3.28 **Dynamic Input** – first example – **Polyline** – *entering* widths

5. *Drag* the cursor to the right until the dimension shows the required horizontal length and *left-click* (Fig. 3.29).

Fig. 3.29 **Dynamic Input** – first example – **Polyline** – the horizontal length

6. *Drag* the cursor down until the vertical distance shows and *left-click* (Fig. 3.30).
7. *Drag* the cursor to the left until the required horizontal distance is showing and *right-click* (Fig. 3.31).

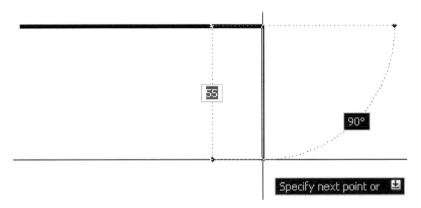

Fig. 3.30 **Dynamic Input** – first example – **Polyline** – the vertical height

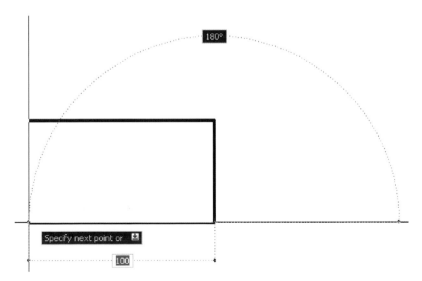

Fig. 3.31 **Dynamic Input** – first example – **Polyline** – the horizontal distance

8. Press the **down** key of the keyboard and *click* **Close** in the menu (Fig. 3.32). The rectangle completes.

Figure 3.33 shows the completed drawing.

DYN – second example – **Zoom**

1. *Click* the **Zoom** icon in the status bar. The first **Zoom** prompt appears (Fig. 3.34).
2. *Right-click* and press the **down** button of the keyboard. The pop-up list (Fig. 3.35) appears from which a **Zoom** prompt can be selected.
3. Carry on using the **Zoom** tool as described in Chapter 4.

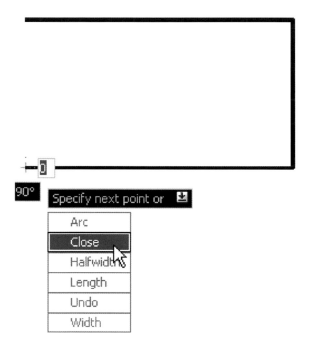

Fig. 3.32 **Dynamic Input** – first example – **Polyline** – selecting **Close** from the pop-up menu

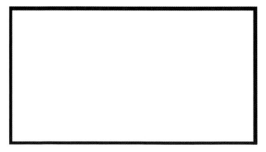

Fig. 3.33 **Dynamic Input** – first example – **Polyline**

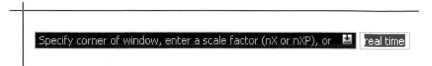

Fig. 3.34 **Dynamic Input** – second example – **Zoom** – click the **Zoom** icon. The prompts then appear

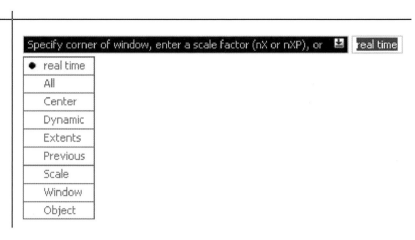

Fig. 3.35 **Dynamic Input** – second example – **Zoom** – the pop-up menu appearing with a *right-click* and pressing the **down** keyboard button

DYN third example – dimensioning

When using **DYN**, tools can equally as well be selected from a toolbar. Figure 3.36 shows the **Linear** tool from the **Dimension** toolbar selected when dimensioning a drawing.

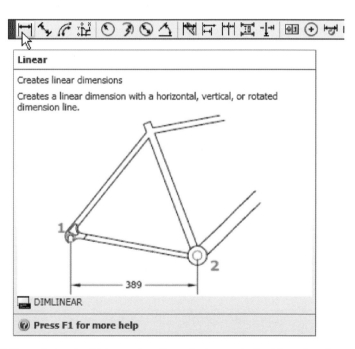

Fig. 3.36 **Dynamic Input** – third example – **dimensioning** – selecting **Linear** from the **Dimension** toolbar

A prompt appears asking for the first point. Move the cursor to the second point; another prompt appears (Fig. 3.37). Press the **down** button of the keyboard and the pop-up list (Fig. 3.37) appears from which a selection can be made.

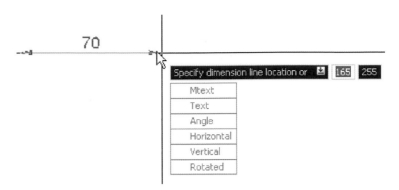

Fig. 3.37 **Dynamic Input** – third example – **dimensioning** – the pop-up menu assopciated with **Linear** dimensioning

Dynamic Input using 3D tools

The **Dynamic Input** method o f constructing 2D drawings can be used equally as well when constructing 3D solid models drawings (see Chapter 4 onwards).

Why use Dynamic Input?

Some operators may prefer constructing drawings without having to make entries at the command line in response to tool prompts. By using **DYN,** drawings – whether in 2D or in 3D format – can be constructed purely from operating and moving the mouse, *entering* coordinates at the command line and pressing the **down** key of the keyboard when necessary.

Examples of using other Draw tools

Polygon tool (Fig. 3.38)

Call the **Polygon** tool – with a *click* on its tool icon in the **Home/Draw** panel (Fig. 3.2, p. 48), from the **Draw** drop-down menu, from the **Draw** toolbar or by *entering* **pol** or **polygon** at the command line. No matter how the tool is called, the command line shows:

```
Command: _polygon Enter number of sides <4>: 6
Specify center of polygon or [Edge]: 60,210
Enter an option [Inscribed in circle/Circumscribed
  about circle] <I>: right-click (accept Inscribed)
Specify radius of circle: 60
Command:
```

2. In the same manner construct a **5**-sided polygon of centre **200,210** and of radius **60**.
3. Then, construct an **8**-sided polygon of centre **330,210** and radius **60**.

4. Repeat to construct a **9**-sided polygon circumscribed about a circle of radius **60** and centre **60,80**.
5. Construct yet another polygon with **10** sides of radius **60** and of centre **200,80**.
6. Finally another polygon circumscribing a circle of radius **60**, of centre **330,80** and sides **12**.

The result is shown in Fig. 3.38.

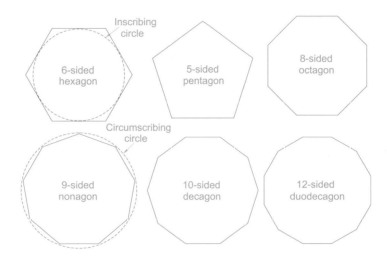

Fig. 3.38 First example – **Polygon** tool

Rectangle tool – First example (Fig. 3.41)

Call the **Rectangle** tool – either with a *click* on its tool icon in the **Home/ Draw** panel (Fig. 3.39) or by *entering* **rec** or **rectangle** at the command line. The tool can be also called from the **Draw** drop-down menu or **Draw** toolbar. The command line shows:

```
Command: _rectang
Specify first corner point or [Chamfer/Elevation/
  Fillet/Thickness/Width]: 25,240
Specify other corner point or [Area/Dimensions/
  Rotation]: 160,160
Command:
```

Rectangle tool – Second example (Fig. 3.41)

```
Command: _rectang
[prompts]: c (Chamfer)
Specify first chamfer distance for rectangles
  <0>: 15
Specify first chamfer distance for rectangles
  <15>: right-click
```

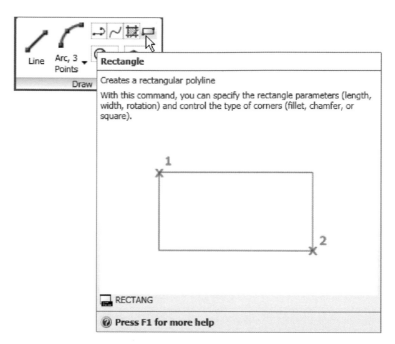

Fig. 3.39 The **Rectangle** tool from the **Home/Draw** panel

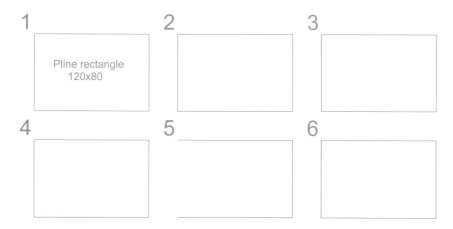

Fig. 3.40 Examples – **Rectangle** Tool

```
Specify first corner point: 200,240
Specify other corner point: 300,160
Command:
```

Rectangle tool – Third example (Fig. 3.41)

```
Command: _rectang
Specify first corner point or [Chamfer/Elevation/
   Fillet/Thickness/Width]: f (Fillet)
Specify fillet radius for rectangles <0>: 15
Specify first corner point or [Chamfer/Elevation/
   Fillet/Thickness/Width]: w (Width)
Specify line width for rectangles <0>: 1
```

```
Specify first corner point or [Chamfer/Elevation/
  Fillet/Thickness/Width]: 20,120
Specify other corner point or [Area/Dimensions/
  Rotation]: 160,30
Command:
```

Rectangle – Fourth example (Fig. 3.41)

```
Command: _rectang
Specify first corner point or [Chamfer/Elevation/
  Fillet/Thickness/Width]: w (Width)
Specify line width for rectangles <0>: 4
Specify first corner point or [Chamfer/Elevation/
  Fillet/Thickness/Width]: c (Chamfer)
Specify first chamfer distance for rectangles
  <0>: 15
Specify second chamfer distance for rectangles
  <15>: right-click
Specify first corner point: 200,120
Specify other corner point: 315,25
Command:
```

The Polyline Edit tool

The **Polyline Edit** tool is a valuable tool for the editing of polylines.

First example – **Polyline Edit** (Fig. 3.43)

1. With the **Polyline** tool construct the outlines **1** to **6** of Fig. 3.40.
2. Call the **Edit Polyline** tool either from the **Home/Modify** panel (Fig. 3.42) or from the **Modify** drop-down menu, or by *entering* **pe** or **pedit** at the command line, which then shows:

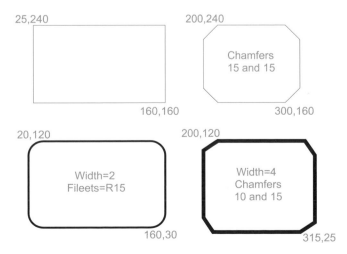

Fig. 3.41 Examples – **Edit Polyline** – the plines to be edited

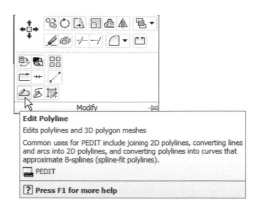

Fig. 3.42 Calling **Edit Polyline** from the **Home**/**Modify** panel

```
Command: enter pe
PEDIT Select polyline or [Multiple]: pick pline 2
Enter an option [Open/Join/Width/Edit vertex/Fit/
  Spline/Decurve/Ltype gen/Undo]: w (Width)
Specify new width for all segments: 2
Enter an option [Open/Join/Width/Edit vertex/Fit/
  Spline/Decurve/Ltype gen/Undo]: right-click
Command:
```

3. Repeat with pline **3** and pedit to Width = **10**.
4. Repeat with line **4** and *enter* **s** (Spline) in response to the prompt line:

```
Enter an option [Open/Join/Width/Edit vertex/
  Fit/Spline/Decurve/Ltype gen/Undo]: enter s
```

5. Repeat with pline **5** and *enter* **j** in response to the prompt line:

```
Enter an option [Open/Join/Width/Edit vertex/
  Fit/Spline/Decurve/Ltype gen/Undo]: enter j
```

The result is shown in pline **6**.
The resulting examples are shown in Fig. 3.43.

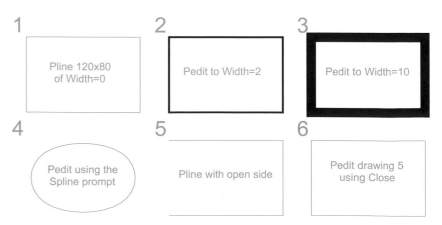

Fig. 3.43 Examples – **Polyline Edit**

Example – Multiple **Polyline Edit** (Fig. 3.44)

1. With the **Polyline** tool construct the left-hand outlines of Fig. 3.44.
2. Call the **Edit Polyline** tool. The command line shows:

```
Command: enter pe
PEDIT Select polyline or [Multiple]: m (Multiple)
Select objects: pick any one of the lines or arcs
  of the left-hand outlines of Fig 6.16 1 found
Select objects: pick another line or arc 1 found
  2 total
```

Continue selecting lines and arcs as shown by the *pick* boxes of the left-hand drawing of Fig. 3.44 until the command line shows:

```
Select objects: pick another line or arc 1 found
  24 total
Select objects: right-click
[prompts]: w (Width)
Specify new width for all segments: 1.5
[prompts]: right-click
Command:
```

The result is shown in the right-hand drawing of Fig. 3.44.

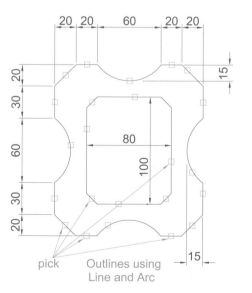
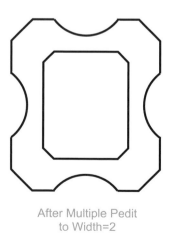

pick   Outlines using
        Line and Arc

After Multiple Pedit
to Width=2

Fig. 3.44 Example – **Multiple Polyline Edit**

Transparent commands

When any tool is in operation it can be interrupted by prefixing the interrupting command with an apostrophe ('). This is particularly useful when wishing to zoom when constructing a drawing (see p. 75). As an example when the **Line** tool is being used:

```
Command: _line
Specify first point: 100,120
Specify next point: 190,120
Specify next point: enter 'z (Zoom)
 >>Specify corner of window or [prompts]: pick
 >>>>Specify opposite corner: pick
Resuming line command.
Specify next point:
```

and so on. The transparent command method can be used with any tool.

The set variable PELLIPSE

Many of the operations performed in AutoCAD are carried out under settings of **set variables**. Some of the numerous set variables available in AutoCAD 2009 will be described in later pages. The variable **PELLIPSE** controls whether ellipses are drawn as splines or as polylines. It is set as follows:

```
Command: enter pellipse right-click
Enter new value for PELLIPSE <0>: enter 1
  right-click
Command:
```

and now when ellipses are drawn they are plines. If the variable is set to **0**, the ellipses will be splines. The value of changing ellipses to plines is that they can then be edited using the **Polyline Edit** tool.

REVISION NOTES

1. The following terms have been used in this chapter:
   - **Field** – a part of a window or of a dialog in which numbers or letters are *entered* or which can be read.
   - **Pop-up list** – a list brought in screen with a *click* on the arrow often found at the right-hand end of a field.
   - **Object** – a part of a drawing which can be treated as a single object. For example, a line constructed with the **Line** tool is an object; a rectangle constructed with the **Polyline** tool is an object; an arc constructed with the **Arc** tool is an object. It will be seen in a later chapter (Chapter 5) that several objects can be formed into a single object.
   - **Toolbar** – a collection of tool icons, all of which have similar functions. For example, in the **Classic AutoCAD** workspace the **Draw** toolbar contains tool icons, those for tools which are used for drawing, and the **Modify** toolbar contains tool icons of those tools used for modifying parts of drawings.
   - **Ribbon** palettes – when working in either of the **2D Drafting & Annotation**, the **Classic AutoCAD** or the **3D Modeling** workspace, tool icons are held panels in the **Ribbon** palettes.

- **Command line** – a line in the command palette which commences with the word **Command**.
- **Snap Mode**, **Grid Display** and **Object Snap** can be toggled with *clicks* on their respective buttons in the status bar. These functions can also be set with function keys – **Snap Mode – F9; Grid Display – F7; Object Snap – F3**.
- **Object Snaps** ensure accurate positioning of objects in drawings.
- **Object Snap** abbreviations can be used at the command line rather than setting in **ON** in the **Drafting Settings** dialog.
- **Dynamic Input** allows constructions in any of the three AutoCAD 2009 workspaces or in a full screen workspace, without having to use the command palette for *entering* the initials of command line prompts.

2. Notes on tools
   - Polygons constructed with the **Polygon** tool are regular polygons – the edges of the polygons are all the same length and the angles are of the same degrees.
   - Polygons constructed with the **Polygon** tool are plines, so can be changed by using the **Edit Polyline** tool.
   - The easiest method of calling the **Edit Polyline** tool is to *enter pe* at the command line.
   - The **Multiple** prompt of the **pedit** tool saves considerable time when editing a number of objects in a drawing.
   - Transparent commands can be used to interrupt tools in operation by preceding the interrupting tool name with an apostrophe (').
   - Ellipses drawn when the variable **PELLIPSE** is set to **0** are splines; when **PELLIPSE** is set to **1**, ellipses are polylines. When ellipses are in polyline form they can be modified using the **pedit** tool.

Exercises

Methods of constructing answers to the following exercises can be found in the free website:

http://books.elsevier.com/companions/9780750689830

1. Using the **Line and Arc** tools, construct the outline given in Fig. 3.45.

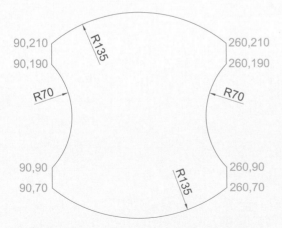

Fig. 3.45 Exercise 1

2. With the **Line** and **Arc** tools, construct the outline shown in Fig. 3.46.

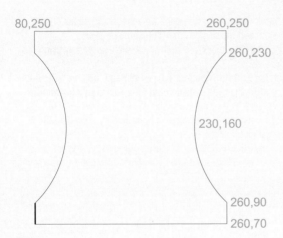

Fig. 3.46 Exercise 2

3. Using the **Ellipse** and **Arc** tools construct the drawing shown in Fig. 3.47.

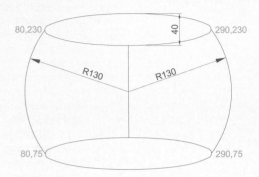

Fig. 3.47 Exercise 3

4. With the **Line**, **Circle** and **Ellipse** tools construct the drawing shown in Fig. 3.48.

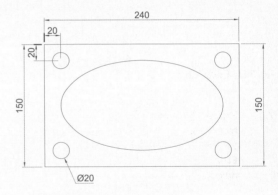

Fig. 3.48 Exercise 4

5. With the **Ellipse** tool, construct the drawing shown in Fig. 3.49.

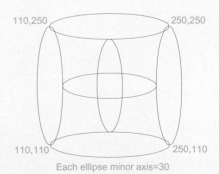

Each ellipse minor axis=30

Fig. 3.49 Exercise 5

6. Figure 3.50 shows a rectangle in the form of a square with hexagons along each edge. Using the **Dimensions** prompt of the **Rectangle** tool construct the square. Then, using the **Edge** prompt of the **Polygon** tool, add the four hexagons. Use the **Object Snap endpoint** to ensure the polygons are in their exact positions.

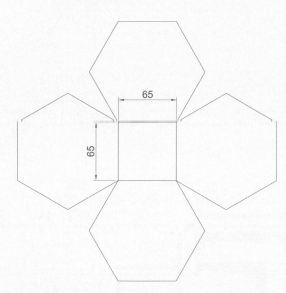

Fig. 3.50 Exercise 6

7. Figure 3.51 shows seven hexagons with edges touching. Construct the inner hexagon using the **Polygon** tool, then with the aid of the **Edge** prompt of the tool, add the other six hexagons.

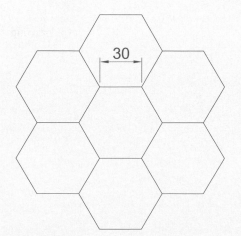

Fig. 3.51 Exercise 7

8. Figure 3.52 was constructed using only the **Rectangle** tool. Make an exact copy of the drawing using only the **Rectangle** tool.

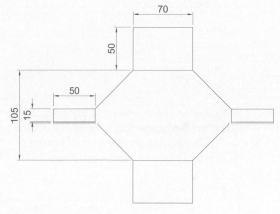

Fig. 3.52 Exercise 8

9. Construct the drawing shown in Figure 3.53 using the **Line** and **Arc** tools. Then, with the aid of the **Multiple** prompt of the **Edit Polyline** tool, change the outlines into plines of **Width = 1**.

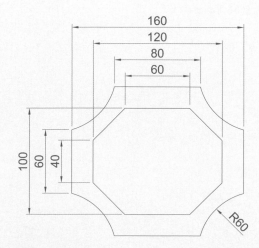

Fig. 3.53 Exercise 9

CHAPTER 3

10. Construct the drawing shown in Fig. 3.54 using the **Line** and **Arc** tools. Then change all widths of lines and arcs to a width of **2** with **Polyline Edit**.

Fig. 3.54 Exercise 10

11. Construct the two outlines shown in Fig. 3.55 using the **Rectangle** and **Line** tools and then with **Edit Polyline** change the parts of the drawing to plines of widths as shown.

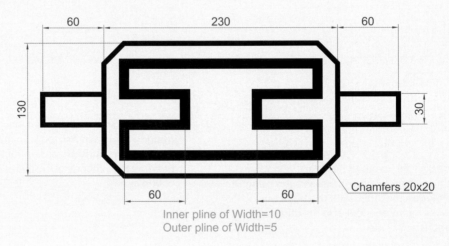

Inner pline of Width=10
Outer pline of Width=5

Chamfers 20x20

Fig. 3.55 Exercise 11

Chapter 4

Zoom, Pan and templates

AIMS OF THIS CHAPTER

The aims of this chapter are:

1. to demonstrate the value of the **Zoom** tools;
2. to introduce the **Pan** tool;
3. to describe the value of using the **Aerial View** window in conjunction with the **Zoom** and **Pan** tools;
4. to update the **acadiso.dwt** template;
5. to describe the construction and saving of drawing templates.

Introduction

The use of the **Zoom** tools allows not only the close inspection of the most minute areas of a drawing in the AutoCAD 2009 drawing area, but allows the accurate construction of very small details in a drawing.

The **Zoom** tools can be called from the **Home/Utilities** panel, from the **Zoom** toolbar or by *clicking* the **Zoom** button in the status bar (Fig. 4.1). However, by far the easiest and quickest method of calling **Zoom** is to *enter* **z** at the command line as follows:

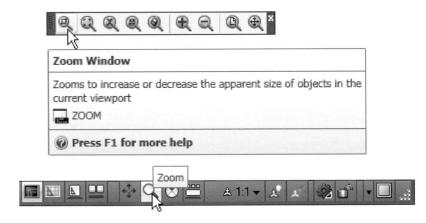

Fig. 4.1 Calling **Zoom Window** from the **Zoom** toolbar or *clicking* the **Zoom** button in the status bar

```
Command: enter z right-click

ZOOM Specify corner of window, enter a scale
factor (nX or nXP) or [All/Center/Dynamic/Extents/
Previous/Scale/Window/Object]<real time>:
```

This allows the different zooms:

- **Realtime** – selects parts of a drawing within a window.
- **All** – the screen reverts to the limits of the template.
- **Center** – the drawing centres itself around a *picked* point.
- **Dynamic** – a broken line surrounds the drawing which can be changed in size and repositioned to part of the drawing.
- **Extents** – the drawing fills the AutoCAD drawing area.
- **Previous** – the screen reverts to its previous zoom.
- **Scale** – entering a number or a decimal fraction scales the drawing.
- **Window** – the parts of the drawing within a *picked* window appear on screen. The effect is the same as using **Realtime**.
- **Object** – *pick* any object on screen and the object zooms.

The operator will probably be using **Realtime**, **Window** and **Previous** zooms most frequently.

Figures 4.2–4.4 show a drawing which has been constructed, a **Zoom Window** of part of the drawing allowing it to be checked for accuracy, and a **Zoom Extents**, respectively.

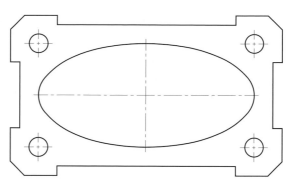

Fig. 4.2 A drawing constructed using the **Polyline** tool

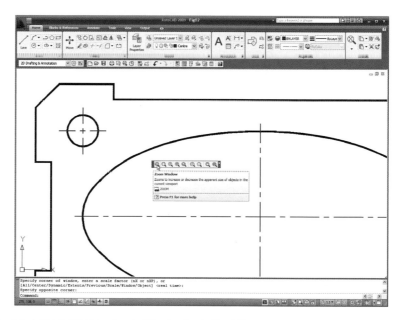

Fig. 4.3 A **Zoom Window** of part of the drawing in Fig. 4.2

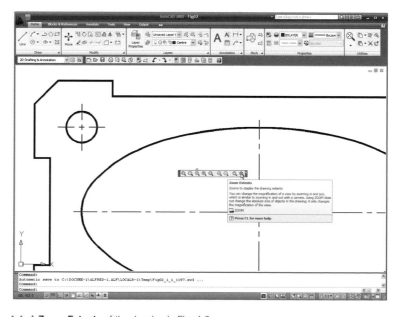

Fig. 4.4 A **Zoom Extents** of the drawing in Fig. 4.2

It will be found that the **Zoom** tools are among those most frequently used when working in AutoCAD 2009.

The **Aerial View** window

Enter **dsviewer** at the command line and the **Aerial View** window appears – usually in the bottom-right-hand corner of the AutoCAD 2009 window. The **Aerial View** window shows the whole of a drawing on screen. The **Aerial View** window is of value when dealing with large drawings – it allows that part of the window on screen to be shown in relation to other parts of the drawing. Figure 4.5 is a three-view orthographic projection of a small bench vice.

Figure 4.6 shows a **Zoom Window** of the drawing in Fig. 4.5, including the **Aerial View** window. The area of the drawing within the **Zoom** window in the drawing area is bounded by a thick black line in the **Aerial View** window.

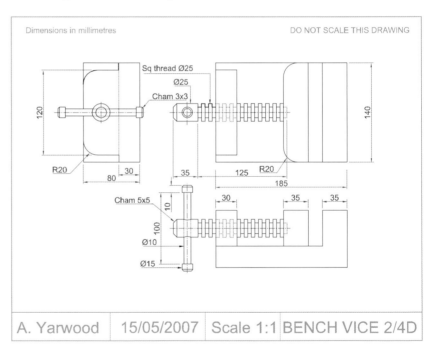

Fig. 4.5 The drawing used to illustrate Figs 4.6 and 4.7

The **Pan** tool

The **Pan** tools can be called with a *click* on the **Pan** icon in the **Home/Utilities** panel, by the **Pan** button in the status bar, from the **Pan** sub-menu of the **View** drop-down menu or by *entering* **p** at the command line. When the tool is called, the cursor on screen changes to an icon of a hand. *Dragging* the hand across the screen under mouse movement allows various parts of the drawing not on screen to be viewed. As the *dragging* takes place, the black rectangle in the **Aerial View** window moves in sympathy (see Fig. 4.7). The **Pan** tool allows any part of the drawing to be viewed and/or modified. When that part of the drawing which is required is on screen a *right-click* calls up the menu as shown in Fig. 4.7, from which either the tool can be exited, or other tools can be called.

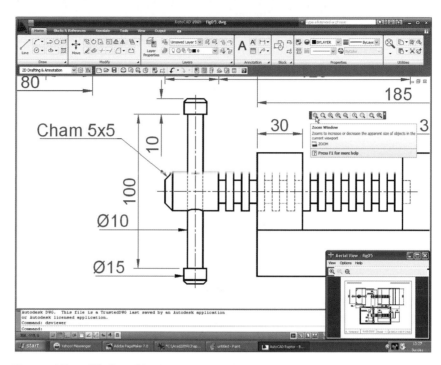

Fig. 4.6 A **Zoom Window** of the drawing Fig. 4.5 with its surrounding zoom rectangle showing in the **Aerial View** window

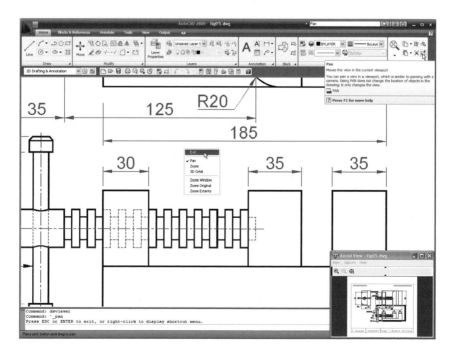

Fig. 4.7 The **Pan** tool in action showing a part of the drawing, while the whole drawing is shown in the **Aerial View** window.

Note

1. If using a mouse with a wheel both zooms and pans can be performed with the aid of the wheel (see p. 8).

2. The **Zoom** tools are important in that they allow even the smallest parts of drawings to be examined and, if necessary, amended or modified.

3. The **Zoom** tools can be called with a *click* on the **Zoom** icon in the **Home/Utilities** panel, with a *click* on the **Zoom** button in the status bar, from the **Zoom** toolbar, from the sub-menu of the **View** drop-down menu or by *entering* **zoom** or **z** at the command line. The easiest of this choice is to *enter* **z** at the command line followed by a *right-click*.

4. Similarly the easiest method of calling the **Pan** tool is to *enter* **p** at the command line followed by a *right-click*.

5. When constructing large drawings, the **Pan** tool and the **Aerial View** window are of value for allowing work to be carried out in any part of a drawing, while showing the whole drawing in the **Aerial View** window.

Drawing templates

In Chapters 1 to 3, drawings were constructed in the template **acadiso.dwt**, which loaded when AutoCAD 2009 was opened. The default **acadiso** template has been amended to **Limits** set to **420,297** (coordinates within which a drawing can be constructed), **Grid Display** set to **10**, **Snap Mode** set to **5**, and the drawing area **Zoomed** to **All**.

Throughout this book most drawings will be based on an **A3** sheet, which measures 420 units by 297 units (the same as **Limits**).

Note

As mentioned on p. 18, if others are using the computer on which drawings are being constructed, it is as well to save the template being used to another file name or, if thought necessary, to a floppy disk. A file name **My_template.dwt**, as suggested earlier, or a name such as **book_template**, can be given.

Adding features to the template

Four other features will now be added to our template:

- **Text style** – set in the **Text Style** dialog.
- **Dimension style** – set in the **Dimension Style Manager** dialog.
- **Shortcutmenu variable** – set to **0**.
- **Layers** – set in the **Layer Properties Manager** dialog.

Setting Text

1. At the command line:

   Command: *enter* st (Style) *right-click*

2. The **Text style** dialog appears (Fig. 4.8). In the dialog, *enter* **6** in the **Height** field. Then *left-click* on **Arial** in the **Font name** popup list. **Arial** font letters appear in the **Preview** area of the dialog.

3. *Left-click* the **New** button and *enter* **Arial** in the **New** text style sub-dialog which appears (Fig. 4.9) and *click* the **OK** button.

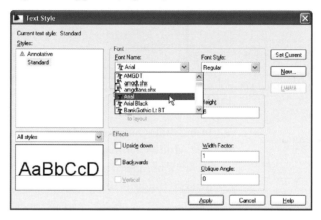

Fig. 4.8 The **Text Style** dialog

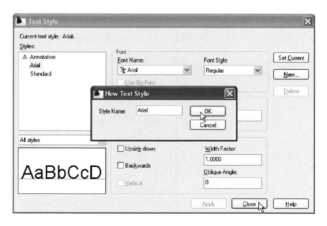

Fig. 4.9 The **New Text Style** sub-dialog

4. *Left-click* the **Set Current** button of the **Text Style** dialog.
5. *Left-click* the **Close** button of the dialog.

Setting dimension style

Settings for dimensions require making *entries* in a number of sub-dialogs in the **Dimension Style Manager**. Set the dimensions style as follows:

1. At the command line:

   Command: *enter* d *right-click*

   and the **Dimensions Style Manager** dialog appears (Fig. 4.10).

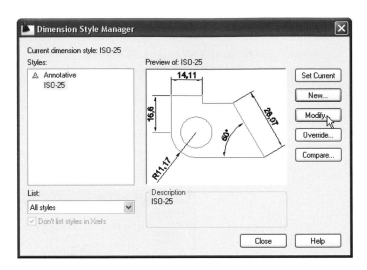

Fig. 4.10 The **Dimensions Style Manager** dialog

2. In the dialog, *click* the **Modify…** button.
3. The **Modify Dimension Style** dialog appears (Fig. 4.11). This dialog shows a number of tabs at the top of the dialog. *Click* the **Lines** tab and make settings as shown in Fig. 4.11. Then *click* the **OK** button of that dialog.

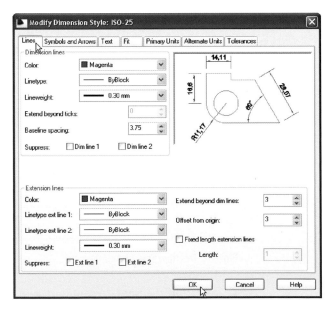

Fig. 4.11 The setting for **Lines** in the **Modify Dimensions Style** dialog

4. The original **Dimension Style Manager** reappears. *Click* its **Modify** button again.
5. The **Modify Dimension Style** dialog reappears (Fig. 4.12), *click* the **Symbols and Arrows** tab. Set **Arrow** size to **6**.
6. Then *click* the **Text** tab. Set **Text style** to **Arial**, set **Color** to **Magenta**, set **Text Height** to **6** and *click* the **ISO** check box in the bottom right-hand corner of the dialog.

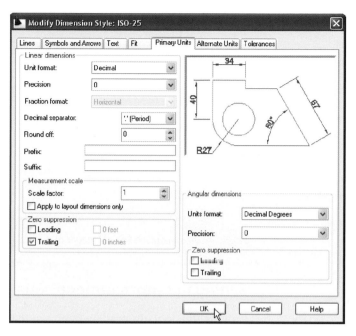

Fig. 4.12 Setting **Primary Units** in the **Dimension Style Manager**

7. Then *click* the **Primary Units** tab and set the units **Precision** to **0**, that is no units after decimal point and **Decimal separator** to **Period**. *Click* the sub-dialogs **OK** button (Fig. 4.12).
8. The **Dimension Styles Manager** dialog reappears showing dimensions, as they will appear in a drawing, in the **Preview of my-style** box. *Click* the **New…** button. The **Create New Dimension Style** dialog appears (Fig. 4.13). *Enter a suitable name in the* **New style name** field – in this example this is **My-style**. *Click* the **Continue** button and the **Dimension Style Manager** appears (Fig. 4.14). This dialog now shows a preview of the **My-style** dimensions. *Click* the dialog's **Set Current** button, following by another *click* on the **Close** button (see Fig. 4.14).

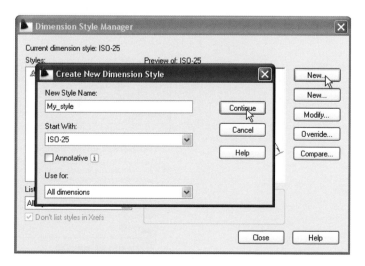

Fig. 4.13 The **Create New Dimension Style** dialog

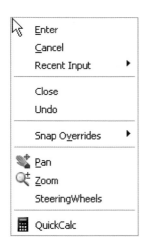

Fig. 4.15 The *right-click* menu

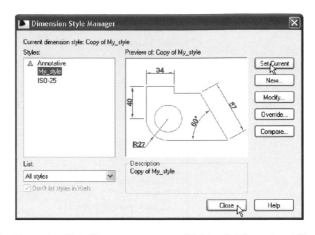

Fig. 4.14 The **Dimension Style Manager** reappears. *Click* the **Set Current** and **Close** buttons

Setting the **shortcutmenu** variable

Call the line tool, draw a few lines and then *right-click*. The *right-click* menu shown in Fig. 4.15 may well appear. The menu will also appear when any tool is called. Some operators prefer using this menu when constructing drawings. To stop this menu appearing:

```
Command: enter shortcutmenu right-click

Enter new value for SHORTCUTMENU <12>: 0

Command:
```

and the menu will no longer appear when a tool is in action.

Setting Layers (see also p. 143)

1. *Left-click* on the **Layer...** tool icon in the **Home/Layers** panel (Fig. 4.16). The **Layer Properties Manager** dialog appears on screen (Fig. 4.17).

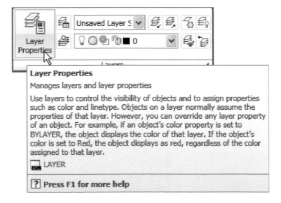

Fig. 4.16 The **Layer Properties Manager** icon in the **Home/Layers** panel

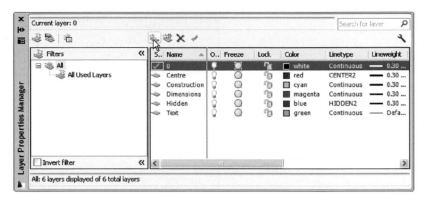

Fig. 4.17 The **Layer Properties Manager** dialog

2. *Click* the **New Layer** icon. **Layer1** appears in the layer list. Overwrite the name **Layer1** *entering* **Center**.
3. Repeat step **2** four times and make four more layers entitled **Construction**, **Dimensions**, **Hidden** and **Text**.
4. *Click* one of the squares under the **Color** column of the dialog. The **Select Color** dialog appears (Fig. 4.18). *Double-click* on one of the colours in the **Index Color** squares. The selected colour appears against the layer name in which the square was selected. Repeat until all 5 new layers have a colour.

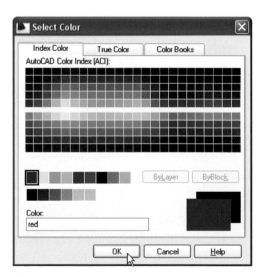

Fig. 4.18 The **Select Color** dialog

5. *Click* on the linetype **Continuous** against the layer name **Center**. The **Select Linetype** dialog appears (Fig. 4.19). *Click* its **Load...** button and from the **Load or Reload Linetypes** dialog *double-click* **CENTER2**. The dialog disappears and the name appears in the **Select Linetype** dialog. *Click* the **OK** button and the linetype **CENTER2** appears against the layer **Center**.
6. Repeat with layer **Hidden**, load the linetype **HIDDEN2** and make the linetype against this layer **HIDDEN2**.

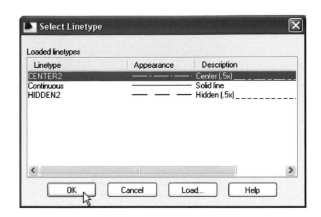

Fig. 4.19 The **Select Linetype** dialog

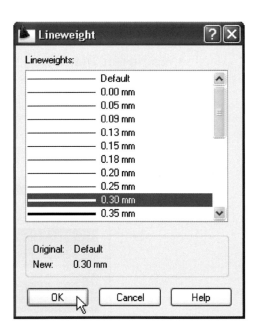

Fig. 4.20 The **Lineweight** dialog

7. *Click* on the any of the lineweights in the **Layer Properties Manager**. This brings up the **Lineweight** dialog (Fig. 4.20). Select the lineweight **0.3**. Repeat the same for all other the layers. Then *click* the **Close** button of the **Layer Properties Manager**.

Saving the template file

1. *Left-click* **Save As…** in the **File** drop-down menu or the **Save** icon in the **Quick Access Toolbar**.
2. In the **Save Drawing As** dialog which comes on screen (Fig. 4.21), *click* the arrow to the right of the **Files of type** field and in the pop-up list associated with the field *click* on **AutoCAD Drawing Template (*.dwt)**. The list of template files in the **AutoCAD 2009/Template** directory appears in the file list.

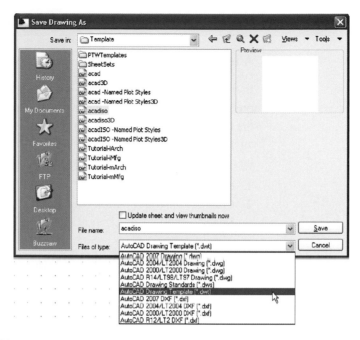

Fig. 4.21 Saving the template to the name **acadiso.dwt**

3. *Click* on **acadiso** in the file list, followed by a *click* on the **Save** button.
4. A **Template Description** dialog appears. Make *entries* as suggested in Fig. 4.22, making sure that **Metric** is chosen from the pop-up list.

Fig. 4.22 The **Template Description** dialog

The template can now saved to be opened for the construction of drawings as needed.

Now when AutoCAD 2009 is opened again the template **acadiso.dwt** appears on screen.

Note

Please remember that if others are using the computer it is advisable to save the template to a name of your own choice.

Another template

A template A3_template.dwt – Fig. 4.24

In the **Select Template** dialog a *click* on any of the file names causes a preview of the template to appear in the **Preview** box of the dialog, unless the template is free of information – as is **acadiso.dwt**. To construct another template which includes a title block and other information based on the **acadiso.dwt** template, proceed as follows:

1. In an **acadiso.dwt** template construct the required border, title block.
2. *Click* the **Layout1** button (Fig. 4.23). The screen changes to a **Paper Space** setting.

Fig. 4.23 The **Model or Paper space** button

3. **Zoom** to **Extents**.
4. It is suggested this template be saved as a **Paper Space** template with the name **A3_template.dwt**.

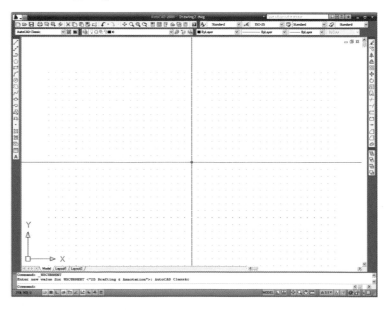

Fig. 4.24 The **A3_template.dwt**

> **Note**
>
> 1. The outline for this template is a pline from **0,290** to **420,290** to **420,0** to **0,0** to **290,0** and of width **0.5**.
>
> 2. The upper line of the title block is a pline from **0,20** to **420,20**.
>
> 3. **Paper Space** is two-dimensional.
>
> 4. Further uses for **Layouts** and **Pspace** are given in Chapter 15.

The AutoCAD Classic workspace

It should be noted that any of the drawings described throughout Part 1: 2D Design of this book could equally well have been constructed in the **AutoCAD Classic** workspace as in the **2D Drafting & Annotation** workspace. Tools could, as well, be selected in the toolbars: **Draw** – *docked* left, **Modify** – *docked* right and **Standard** – *docked* top, as can be seen in Fig. 4.25.

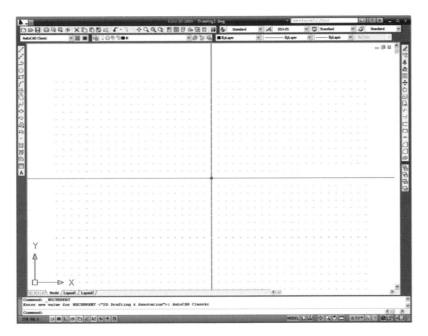

Fig. 4.25 The **AutoCAD Classic** workspace

REVISION NOTES

1. The **Zoom** tools are important in that they allow even the smallest parts of drawings to be examined and, if necessary, amended or modified.
2. The **Zoom** tools can be called with a *click* on the **Zoom** icon in the **Home/Utilities** panel, *clicking* the **Zoom** button in the status bar, from the **Zoom** toolbar, from the sub-menu of the **View** drop-down menu, or by *entering* **z** or **zoom** at the command line. The easiest is to *enter* **z** at the command line followed by a *right-click*.
3. There are five methods of calling tools for use: selecting a tool icon in a panel from a group of panels shown as **tabs** in the **Ribbon**; selecting a tool from a toolbar; *entering* the name of a tool in full at the command line; *entering* an abbreviation for a tool; selecting a tool from a drop-down menu.
4. When constructing large drawings, the **Pan** tool and the **Aerial View** window are of value for allowing work to be carried out in any part of a drawing, while showing the whole drawing in the **Aerial View** window.
5. An A3 sheet of paper is 420 mm by 297 mm. If a drawing constructed in the template **acadiso.dwt** described in this book, is printed/plotted full size (scale 1:1), each unit in the drawing will be 1 mm in the print/plot.
6. When limits are set it is essential to call **Zoom** followed by **a** (All) to ensure that the limits of the drawing area are as set.
7. If the *right-click* menu appears when using tools, the menu can be aborted if required by setting the **SHORTCUTMENU** variable to **0**.

CHAPTER 4

Exercises

If you have saved drawings constructed either by following the worked examples in this book or by answering exercises in Chapters 2 and 3, open some of them and practise zooms and pans.

Chapter 5

The Modify tools

AIM OF THIS CHAPTER

The aim of this chapter is to describe the uses of tools for modifying parts of drawings.

Introduction

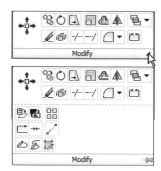

The **Modify** tools are among those most frequently used. The tools are found in the **Home/Modify** panel. A *click* on the arrow in the **Home/Modify** panel brings down a further set of tool icons (Fig. 5.1). They can also be selected from the **Modify** toolbar (Fig. 5.2) or from the **Modify** drop-down menu.

Using the **Erase** tool from **Home/Modify** was described in Chapter 2. Examples of tools other than the **Explode** follow. See also Chapter 10 for **Explode**.

Fig. 5.1 The **Modify** tool icons in the **Home/Modify** panel

Fig. 5.2 The **Modify** toolbar

The **Copy** tool

First example – **Copy** (Fig. 5.5)

1. Construct Fig. 5.3 using **Polyline**. Do not include the dimensions.
2. Call the **Copy** tool – *left-click* on its tool icon in the **Home/Modify** panel (Fig. 5.4), *pick* **Copy** from the **Modify** toolbar, or *enter* **cp** or **copy** at the command line.

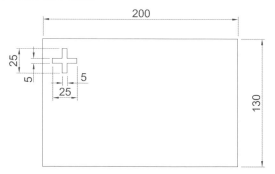

Fig. 5.3 First example – **Copy Object** – outlines

The command line shows:

```
Command: _copy
Select objects: pick the cross 1 found
Select objects: right-click
Current settings: Copy mode = Multiple
Specify base point or [Displacement/mode/Multiple]
  <Displacement>: pick
Specify second point or <use first point as
  displacement>: pick
```

The result is given in Fig. 5.5.

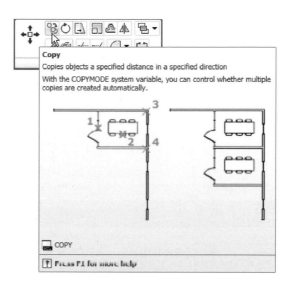

Fig. 5.4 The **Copy** tool from the **Home/Modify** panel

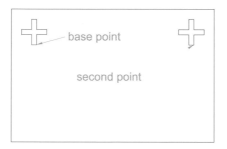

Fig. 5.5 First example – **Copy**

Second example – Copy – Multiple copy (Fig. 5.6)

1. Erase the copied object.
2. Call the **Copy** tool. The command line shows:

```
Command: _copy
Select objects: pick the cross 1 found
Select objects: right-click
Current settings: Copy mode = Multiple
Specify base point or [Displacement/mode]
  <Displacement >: pick
Specify second point or <use first point as
  displacement >: pick
Specify second point or [Exit/Undo] <Exit>: pick
Specify second point or [Exit/Undo] <Exit>: pick
Specify second point or [Exit/Undo] <Exit>:
  e (Exit)
Command:
```

The result is shown in Fig. 5.6.

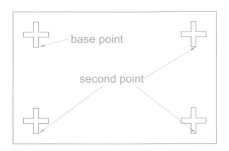

Fig. 5.6 Second example – **Copy** – **Mutiple** – **Multiple copy**

The **Mirror** tool

First example – **Mirror** (Fig. 5.9)

1. Construct the outline shown in Fig. 5.8 using the **Line** and **Arc** tools.
2. Call the **Mirror** tool – *left-click* on its tool icon in the **Home/Modify** panel (Fig. 5.7), *pick* the **Mirror** tool icon from the **Modify** toolbar, *pick* **Mirror** from the **Modify** drop-down menu, or *enter* **mi** or **mirror** at the command line. The command line shows:

```
Command: _mirror
Select objects: pick first corner Specify opposite
  corner: pick 7 found
Select objects: right-click
Specify first point of mirror line: end of pick
Specify second point of mirror line: end of pick
Erase source objects [Yes/No] <N >: right-click
Command:
```

The result is shown in Fig. 5.9.

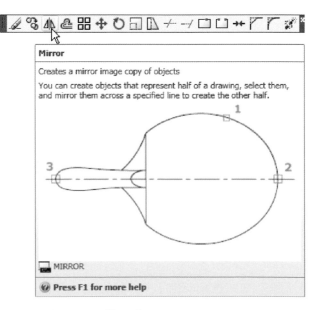

Fig. 5.7 The **Mirror** tool from the **Modify** toolbar

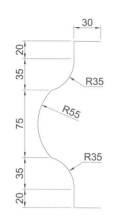

Fig. 5.8 First example –
Mirror – outline

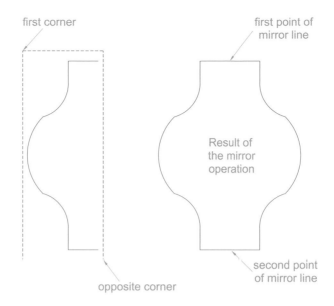

Fig. 5.9 First example – **Mirror**

Second example – **Mirror** (Fig. 5.10)

1. Construct the outline shown in the dimensioned polyline in the upper
   drawing of Fig. 5.10.
2. Call **Mirror** and using the tool three times complete the given outline.
   The two points shown in Fig. 5.10 are to mirror the right-hand side of
   the outline.

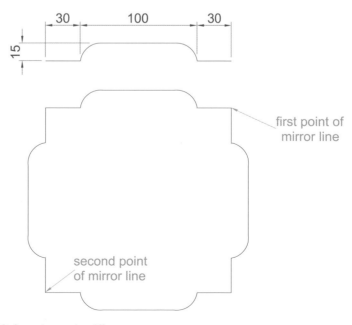

Fig. 5.10 Second example – **Mirror**

Fig. 5.11 Third example – **Mirror**

Third example – **Mirror** (Fig. 5.11)

If text is involved when using the **Mirror** tool, the set variable **MIRRTEXT** must be set correctly. To set the variable:

```
Command: mirrtext
Enter new value for MIRRTEXT <1>: 0
Command:
```

If set to **0** text will mirror without distortion. If set to **1** text will read backwards as indicated in Fig. 5.11.

The **Offset** tool

Examples – **Offset** (Fig. 5.14)

1. Construct the four outlines shown in Fig. 5.13.
2. Call the **Offset** tool – *left-click* its tool icon in the **Home/Modify** panel (Fig. 5.12), *pick* the tool from the **Modify** toolbar, *pick* the tool name in the **Modify** drop-down menu, or *enter* **o** or **offset** at the command line. The command line shows:

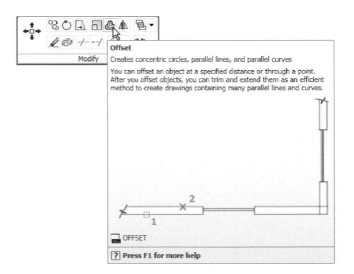

Fig. 5.12 The **Offset** tool from the **Home/Modify** panel

```
Command: _offset
Current settings: Erase source = No Layer =
  Source OFFSETGAPTYPE = 0
Specify offset distance or [Through/Erase/Layer]
  <Through>: 10
Select object to offset or [Exit/Undo] <Exit>:
  pick drawing 1
Specify point on side to offset or [Exit/Multiple/
  Undo] <Exit>: pick inside the rectangle
```

```
Select object to offset or [Exit/Undo] <Exit>:
  e (Exit)
Command:
```

3. Repeat for drawings **2**, **3** and **4** in Fig. 5.13 as shown in Fig. 5.14.

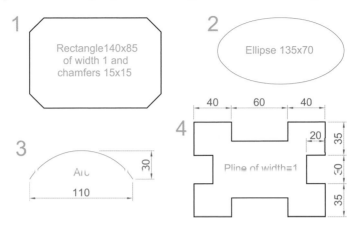

Fig. 5.13 Examples – **Offset** – outlines

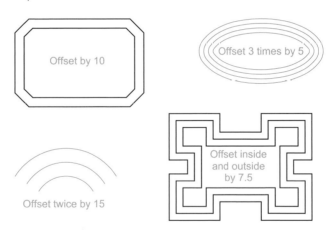

Fig. 5.14 Examples – **Offset**

The Array tool

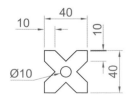

Fig. 5.15 First example – **Array** – drawing to be arrayed

Arrays can be either in a **Rectangular** form or in a **Polar** form as shown in the examples below.

First example – **Rectangular Array** (Fig. 5.17)

1. Construct the drawing as shown in Fig. 5.15.
2. Call the **Array** tool – either *click* **Array** in the **Modify** drop-down menu (Fig. 5.16), from the **Home/Modify** panel, *pick* the **Array** tool icon from the **Modify** toolbar, or *enter* **ar** or **array** at the command line. The **Array** dialog appears (Fig. 5.17).
3. Make settings in the dialog:
  - **Rectangular Array** radio button set on (dot in button)
  - **Row** field – *enter* **5**

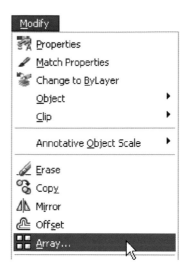

Fig. 5.16 Selecting **Array** from the **Modify** drop-down menu

- **Column** field – *enter* **6**
- **Row offset** field – *enter* **− 50** (note the minus sign)
- **Column offset** field – *enter* **50**.

4. *Click* the **Select objects** button and the dialog disappears. Window the drawing. A second dialog appears which includes a **Preview** < button.

5. *Click* the **Preview** < button. The dialog disappears and the following prompt appears at the command line:

   ```
   Pick or press Esc to return to drawing or
       <Right-click to accept drawing>:
   ```

6. If satisfied *right-click*. If not, press the **Esc** key and make revisions to the **Array** dialog fields as necessary.

The resulting array is shown in Fig. 5.18.

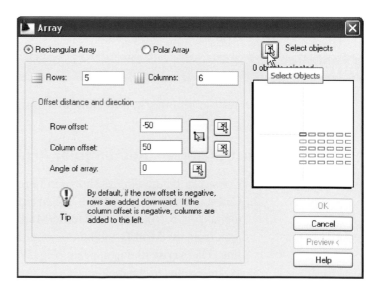

Fig. 5.17 First example – the **Array** dialog

Fig. 5.18 First example – **Array**

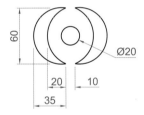

Fig. 5.19 Second example –
the setting in the **Array**
dialog

Second example – **Polar Array** (Fig. 5.22)

1. Construct the drawing shown in Fig. 5.19.
2. Call **Array**. The **Array** dialog appears. Make settings as shown in Fig. 5.20.

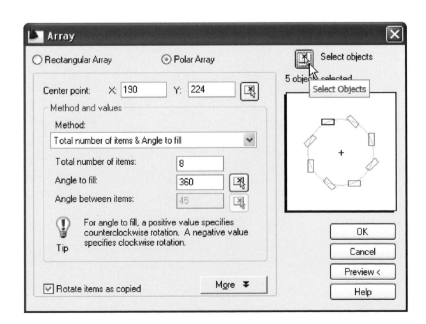

Fig. 5.20 Second example – **Array** – settings in the dialog

3. *Click* the **Select objects** button of the dialog and window the drawing. The dialog returns to screen. *Click* the **Pick center point** button (Fig. 5.21) and when the dialog disappears, *pick* a center point for the array.

4. The dialog reappears. *Click* its **Preview** < button. The array appears and the command line shows:

```
Pick or press Esc to return to drawing or
   <Right-click to accept drawing>:
```

5. If satisfied *right-click* If not, press the **Esc** key and make revisions to the **Array** dialog fields as necessary.

The resulting array is shown in Fig. 5.22.

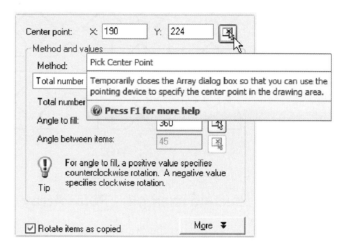

Fig. 5.21 Second example – **Array** – the **Pick Center point** button

Fig. 5.22 Second example – **Array**

The **Move** tool

Example – **Move** (Fig. 5.25)

1. Construct the drawing shown in Fig. 5.23.
2. Call **Move** – *click* the **Move** tool icon in the **Home/Modify** toolbar (Fig. 5.24), *pick* **Move** from the **Modify** panel, *pick* **Move** from the **Modify** drop-down menu, or *enter* **m** or **move** at the command line, which shows:

```
Command: _move
Select objects: pick .the middle shape in the
  drawing 1 found
Select objects: right-click
Specify base point or [Displacement]
  <Displacement>: pick
Specify second point or <use first point as
  displacement>: pick
Command:
```

The result is given in Fig. 5.25.

Rectangle 190x50. Chamfers 10x10

All edges are 5

Fig. 5.23 Example – **Move** – drawing

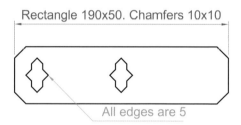

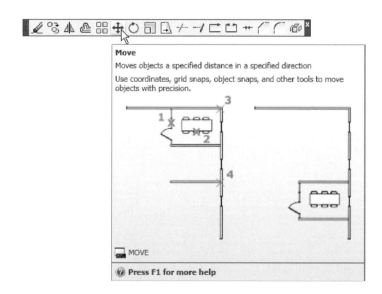

Move

Moves objects a specified distance in a specified direction

Use coordinates, grid snaps, object snaps, and other tools to move objects with precision.

MOVE

ⓘ **Press F1 for more help**

Fig. 5.24 The **Move** tool from the **Home/Modify** toolbar

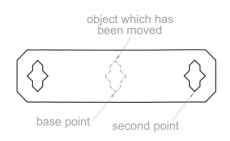

object which has
been moved

base point   second point

Fig. 5.25 Example – **Move**

The **Rotate** tool

When using the **Rotate** tool remember the default rotation of objects within AutoCAD 2009 is counterclockwise (anticlockwise).

Example – **Rotate** (Fig. 5.27)

1. Construct drawing **1** of Fig. 5.27 with **Polyline**. Copy the drawing **1** three times (Fig. 5.27).
2. Call **Rotate** – *left-click* its tool icon in the **Home/Modify** panel (Fig. 5.26), *pick* its tool icon from the **Modify** toolbar, *pick* **Rotate** from the **Modify** drop-down menu, or *enter* **ro** or **rotate** at the command line. The command line shows:

```
Command: _rotate
Current positive angle in UCS: ANGDIR =
  counterclockwise ANGBASE = 0
Select objects: window the drawing 3 found
Select objects: right-click
Specify base point: pick
```

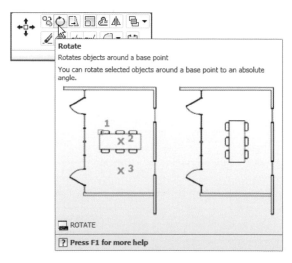

Fig. 5.26 The **Rotate** tool icon from the **Home/Modify** panel

```
Specify rotation angle or [Copy/Reference] <0>: 45
Command:
```

and the first copy rotates through the specified angle.

3. Repeat for drawings **3** and **4** rotating as shown in Fig. 5.27.

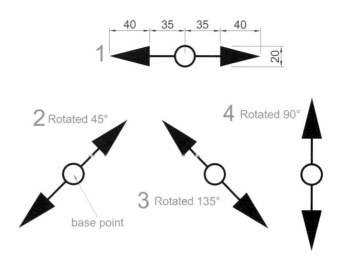

Fig. 5.27 Example **Rotate**

The **Scale** tool

Examples – **Scale** (Fig. 5.29)

1. Using the **Rectangle and** Polyline tools, construct drawing **1** of Fig. 5.29. The **Rectangle** fillets are R10. The line width of all parts is **1**. Copy the drawing 3 times to give drawings **2**, **3** and **4**.
2. Call **Scale** – *left-click* its tool icon in the **Home/Draw** panel (Fig. 5.28), *pick* its tool icon in the **Modify** toolbar, *pick* **Scale** from the **Modify** drop-down-menu or, *click* its tool in the **Modify** toolbar or, *enter* **sc** or **scale** at the command line which then shows:

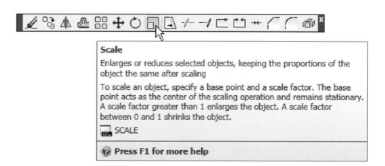

Fig. 5.28 The **Scale** tool icon from the **Home/Modify** toolbar

```
Command: _scale
Select objects: window drawing 2 5 found
Select objects: right-click
Specify base point: pick
Specify scale factor or [Copy/Reference] <1>: 0.75
Command:
```

3. Repeat for the other two drawings **3** and **4** scaling to the scales given with the drawings.

The results are shown in Fig. 5.29.

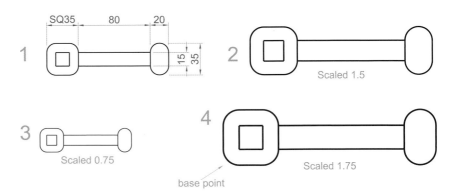

Fig. 5.29 Examples – **Scale**

The **Trim tool**

This tool is one which will be in frequent use when constructing drawings.

First example – **Trim** (Fig. 5.31)

1. Construct the drawing **Original drawing** shown in Fig. 5.31.
2. Call **Trim** – *left-click* its tool icon in the **Home/Modify** panel, *pick* its tool icon in the **Modify** toolbar Fig. 5.30), *pick* **Trim** from the **Modify** drop-down menu, or *enter* **tr** or **trim** at the command line, which then shows:

```
Command: _trim
Current settings: Projection UCS. Edge = None
Select cutting edges: pick the left-hand circle 1
  found
Select objects: right-click
Select objects to trim or shift-select to extend
  or [Fence/Project/Crossing/Edge/eRase//Undo]:
  pick one of the objects
Select objects to trim or shift-select to extend or
```

```
[Fence/Crossing/Project/Edge/eRase/Undo]: pick the
  second of the objects
Select objects to trim or shift-select to extend
  or [Project/Edge/Undo]: right-click
Command:
```

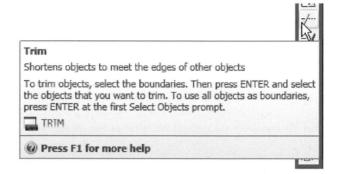

Fig. 5.30 The **Trim** tool icon from the **Modify** toolbar in the **AutoCAD Classic** workspace

3. This completes the **First stage** as shown in Fig. 5.31. Repeat the **Trim** sequence for the **Second stage**.
4. The **Third stage** drawing of Fig. 5.31 shows the result of the trims at the left-hand end of the drawing.
5. Repeat for the right-hand end. The final result is shown in the drawing labelled **Result** in Fig. 5.31.

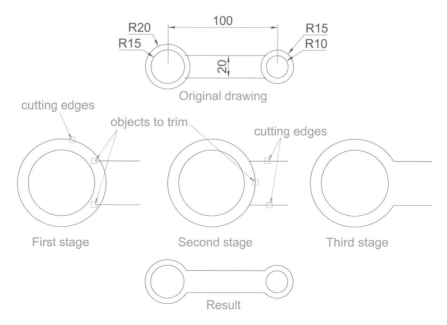

Fig. 5.31 First example – **Trim**

CHAPTER 5

Second example – **Trim** (Fig. 5.32)

1. Construct the left-hand drawing of Fig. 5.32.
2. Call **Trim**. The command line shows:

```
Command: _trim
Current settings: Projection UCS. Edge = None
Select cutting edges…
Select objects or <select all>: pick the
  left-hand arc 1 found
Select objects: right-click
Select objects to trim or shift-select to extend or
  [Fence/Crossing/Project/Edge/eRase/Undo]: e (Edge)
Enter an implied edge extension mode [Extend/No
  extend] <No extend>: e (Extend)
Select objects to trim: pick
Select objects to trim: pick
Select objects to trim: right-click
Command:
```

3. Repeat for the other required trims. The result is given in Fig. 5.32.

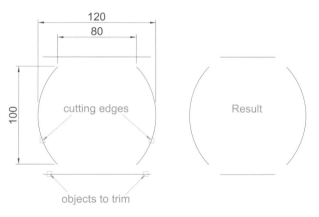

Fig. 5.32 Second example – **Trim**

The **Stretch** tool

Examples – **Stretch** (Fig. 5.34)

As its name implies, the **Stretch** tool is for stretching drawings or parts of drawings. The action of the tool prevents it from altering the shape of circles in any way. Only **crossing** or **polygonal** windows can be used to determine the part of a drawing which is to be stretched.

1. Construct the drawing labelled **Original** in Fig. 5.34, but do not include the dimensions. Use the **Circle**, **Arc**, **Trim** and **Polyline Edit** tools. The resulting outlines are plines of width = 1. With the **Copy** tool make two copies of the drawing.

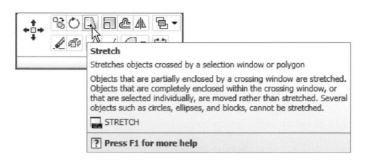

Fig. 5.33 The **Stretch** tool icon from the **Home/Modify** panel

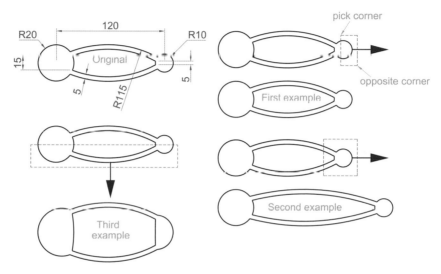

Fig. 5.34 Examples – **Stretch**

Note

In each of the three examples in Fig. 5.34, the broken lines represent the crossing windows required when **Stretch** is used.

---

2. Call the **Stretch** tool – *click* on its tool icon in the **Home/Modify** panel (Fig. 5.33), *left-click* on its tool icon in the **Modify** toolbar, *pick* its name in the **Modify** drop-down menu or *enter* **s** or **stretch** at the command line, which shows:

```
Command: _stretch
Select objects to stretch by crossing-window
  or crossing-polygon...
Select objects: enter c right-click
Specify first corner: pick Specify opposite
  corner: pick 1 found
Select objects: right-click
Specify base point or [Displacement],
  <Displacement>: pick beginning of arrow
```

CHAPTER 5

```
Specify second point of displacement or <use first
point as displacement>: drag in the direction
of the arrow to the required second point and
right-click
Command:
```

> **Note**
>
> 1. When circles are windowed with the crossing window no stretching can take place. This is why, in the case of the first example in Fig. 5.33, when the **second point of displacement** was *picked*, there was no result – the outline did not stretch.
>
> 2. Care must be taken when using this tool as unwanted stretching can occur.

The **Break** tool

Examples – **Break** (Fig. 5.36)

1. Construct the rectangle, arc and circle (Fig. 5.36).
2. Call **Break** – *click* its tool icon in the **Home/Modify** panel (Fig. 5.35), *pick* its tool icon in the **Modify** toolbar, *click* **Break** in the **Modify** drop-down menu, or *enter* **br** or **break** at the command line, which shows:

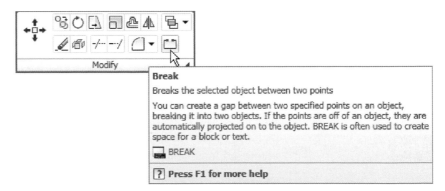

Fig. 5.35 The **Break** tool icon from the **Home/Modify** panel

For drawings 1 and 2

```
Command: _break Select object: pick at the point
Specify second break point or [First point]: pick
Command:
```

For drawing 3

```
Command: _break Select object pick at the point
Specify second break point or [First point]: enter
  f right-click
Specify first break point: pick
Specify second break point: pick
Command:
```

The results are shown in Fig. 5.36.

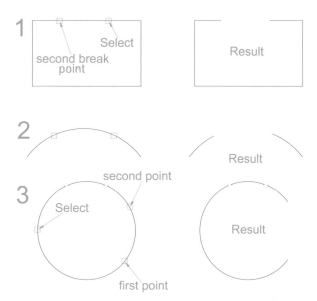

Fig. 5.36 Examples – **Break**

> **Note**
>
> Remember the default rotation of AutoCAD 2009 is counterclockwise. This applies to the use of the **Break** tool.

The **Join** tool

The **Join** tool can be used to join plines provided their ends are touching; to join lines which are in line with each other; to join arcs and convert arcs to circles.

Examples – **Join** (Fig. 5.38)

1. Construct a rectangle from four separate plines – drawing **1** of Fig. 5.38; construct two lines – drawing **2** of Fig. 5.38 and an arc – drawing **3** of Fig. 5.38.

CHAPTER 5

2. Call the **Join** tool – *click* the **Join** tool icon in the **Home/Modify** panel (Fig. 5.37), *left-click* its tool icon in the **Modify** toolbar, select **Join** from the **Modify** drop-down menu or *enter* **join** or **j** at the command line. The command line shows:

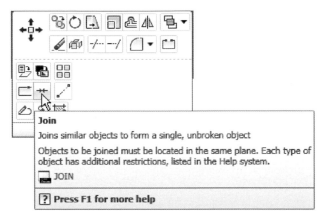

Fig. 5.37   The **Join** tool icon from the **Home/Modify** panel

```
Command: _join Select source object:
Select objects to join to source: pick a pline
  1 found
Select objects to join to source: pick another
  1 found, 2 total
Select objects to join to source: pick another
  1 found, 3 total
Select objects to join to source: right-click
3 segments added to polyline
Command: right-click
JOIN Select source object: pick one of the
  lines
Select lines to join to source: pick the other
  1 found
Select lines to join to source: right-click
  1 line joined to source
Command: right-click
JOIN Select source object: pick the arc
Select arcs to join to source or [close]: enter
  1 right-click
Arc converted to a circle.
Command:
```

The results are shown in Fig. 5.38.

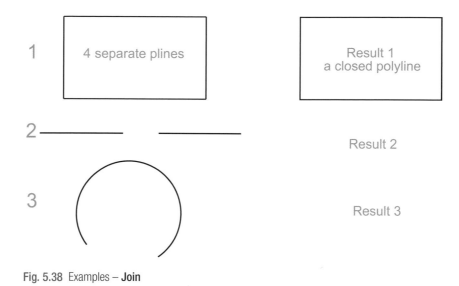

Fig. 5.38 Examples – **Join**

The **Extend** tool

Examples – **Extend** (Fig. 5.40)

1. Construct plines and a circle as shown in the left-hand drawings of Fig. 5.40.
2. Call **Extend** – *click* the **Extend** tool in the **Home/Draw** panel, *click* its tool icon in the **Modify** toolbar (Fig. 5.39), *pick* **Extend** from the **Modify** drop-down menu, or *enter* **ex** or **extend** at the command line which then shows:

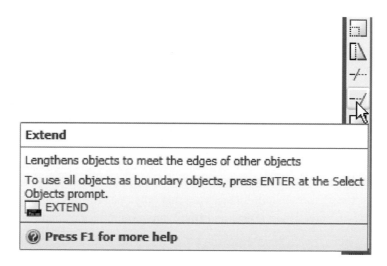

Fig. 5.39 The **Extend** tool icon from the **Modify** toolbar in the **AutoCAD Classic** workspace

CHAPTER 5

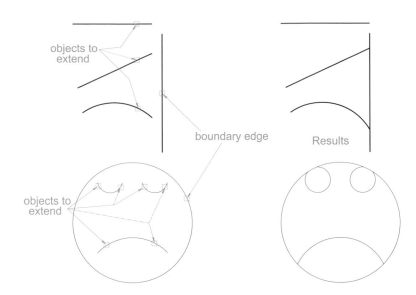

Fig. 5.40 Examples – **Extend**

```
Command: _extend
Current settings: Projection = UCS Edge = Extend
Select boundary edges...
Select objects or <select all >: pick 1 found
Select objects: right-click
Select object to extend or shift-select to trim
  or[Fence/Crossing/Project/Edge/Undo]: pick
```

Repeat for each object to be extended. Then:

```
Select object to extend or shift-select to trim or
  [Fence/Crossing/Project/Edge/Undo]: right-click
Command:
```

Note

Observe the similarity of the **Extend** and **No extend** prompts with those of the **Trim** tool.

The **Fillet** and **Chamfer** tools

These two tools can be called from the **Home/Modify** panel. There are similarities in the prompt sequences for these two tools. The major differences are that only one (**Radius**) setting is required for a fillet, but two (**Dist1** and **Dist2**) are required for a chamfer. The basic prompts for both are:

Fillet

```
Command: _fillet
Current settings: Mode = TRIM, Radius = 1
```

```
Select first object or [Polyline/Radius/Trim/
  multiple]: enter r (Radius) right-click
Specify fillet radius <1>: 15
```

Chamfer

```
Command: _chamfer
(TRIM mode) Current chamfer Dist1 = 1, Dist2 = 1
Select first line or [Undo/Polyline/Distance/Angle/
  Trim/method/Multiple]: enter d (Distance)
  right-click
Specify first chamfer distance <1>: 10
Specify second chamfer distance <10>: right-click
```

Examples – Fillet (Fig. 5.42)

1. Construct three rectangles 100 by 60 using either the **Line** or the **Polyline** tool (Fig. 5.42).
2. **Call** Fillet – *click* the arrow to the right of the tool icon in the **Home/ Modify** panel and select **Fillet** from the menu which appears (Fig. 5.41), *pick* its tool icon in the **Modify** toolbar, *pick* **Fillet** from the **Modify** drop-down menu, or *enter* **f** or **fillet** at the command line which then shows:

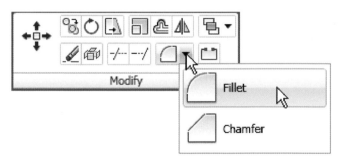

Fig. 5.41 Select **Fillet** from the menu in the **Home/Modify** panel

```
Command: _fillet
Current settings: Mode = TRIM, Radius = 1
Select first object or [Polyline/Radius/Trim/
  Multiple]: r (Radius)
Specify fillet radius <0>: 15
Select first object or [Undo/Polyline/Radius/Trim/
  Multiple]: pick
Select second object or shift-select to apply
  corner: pick
Command:
```

Three examples are given in Fig. 5.42.

CHAPTER 5

CHAPTER 5

Fig. 5.42 Examples – **Fillet**

Examples – **Chamfer** (Fig. 5.44)

1. Construct three rectangles 100 by 60 using either the **Line** or the **Polyline** tool.
2. Call **Chamfer** – *click* the arrow to the right of the tool icon in the **Home/Modify** panel and select **Chamfer** from the menu which appears (Fig. 5.43), *click* on its tool icon in the **Modify** toolbar, *pick* **Chamfer** from the **Modify** drop-down menu, or *enter* **cha** or **chamfer** at the command line which then shows:

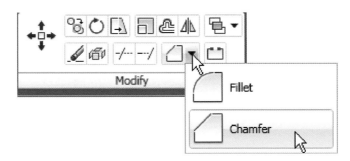

Fig. 5.43 Select **Chamfer** from the **Modify** panel

```
Command: _chamfer
(TRIM mode) Current chamfer Dist1 = 1, Dist2 = 1
Select first line or [Undo/Polyline/Distance/
  Angle/Trim/
Method/Multiple]: d
Specify first chamfer distance <1>: 10
Specify second chamfer distance <10>: right-click
Select first line or [Undo/Polyline/Distance/Angle/
  Trim/Method/Multiple]: pick the first line for
  the chamfer
```

CHAPTER 5

```
Select second line or shift-select to apply
  corner: pick
Command:
```

The other two rectangles are chamfered in a similar manner except that the **No trim** prompt is brought into operation with the bottom left-hand example.

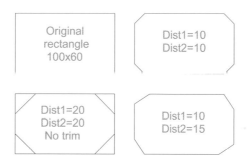

Fig. 5.44 Examples – **Chamfer**

CHAPTER 5

5. With **Offset** the **Through** prompt can be answered by *clicking* two points in the drawing area the distance of the desired offset distance.
6. **Polar Arrays** can be arrays around any angle set in the **Angle of array** field of the **Array** dialog.
7. When using **Scale**, it is advisable to practise the **Reference** prompt.
8. The **Trim** tool in either its **Trim** or its **No trim** modes is among the most useful tools in AutoCAD 2009.
9. When using **Stretch**, circles are unaffected by the stretching.

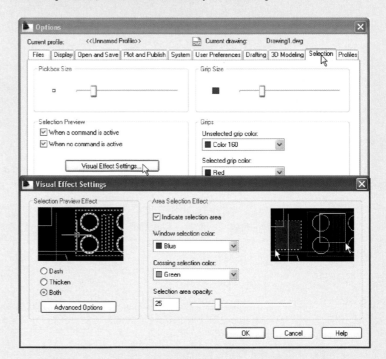

Fig. 5.45 **Visual Setting Effects Settings** sub-dialog of the **Options** dialog

Exercises

Methods of constructing answers to the following exercises can be found in the free website:

http://books.elsevier.com/companions/9780750689830

1. Construct the drawing given in Fig. 5.46. All parts are plines of width = 0.7 with corners filleted R10. The long strips have been constructed using **Circle**, **Polyline**, **Trim** and **Polyline Edit**. Construct one strip and then copy it using **Copy**.

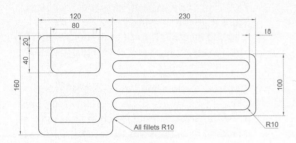

Fig. 5.46 Exercise 1

2. Construct the drawing given in Fig. 5.47. All parts of the drawing are plines of width = 0.7. The setting in the **Array** dialog is to be **180** in the **Angle of array** field.

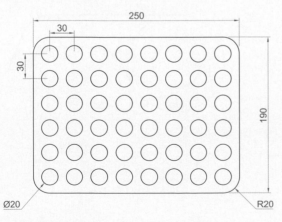

Fig. 5.47 Exercise 2

3. Using the tools **Polyline**, **Circle**, **Trim**, **Polyline Edit**, **Mirror** and **Fillet** construct the drawing given in Fig. 5.48.

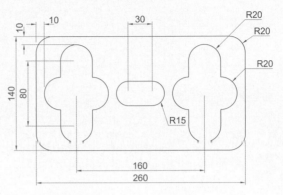

Fig. 5.48 Exercise 3

4. Construct the circles and lines shown in Fig. 5.49. Using **Offset** and the **Ttr** prompt of the **Circle** tool followed by **Trim**, construct one of the outlines arrayed within the outer circle. Then, with **Polyline Edit**, change the lines and arcs into a pline of width = 0.3. Finally, array the outline twelve times around the centre of the circles (Fig. 5.50).

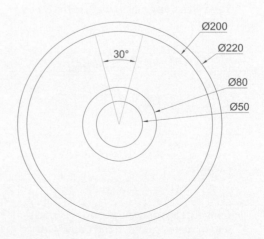

Fig. 5.49 Exercise 4 – circles and lines on which the exercise is based

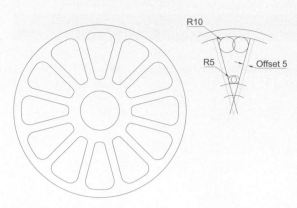

Fig. 5.50 Exercise 4

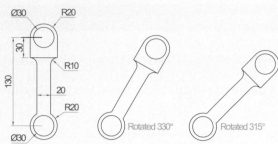

Fig. 5.53 Exercise 7

5. Construct the arrow shown in Fig. 5.51. Array the arrow around the centre of its circle 8 times to produce the right-hand drawing of Fig. 5.51.

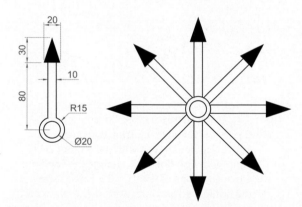

Fig. 5.51 Exercise 5

6. Construct the left-hand drawing of Fig. 5.52. Then with **Move**, move the central outline to the top left-hand corner of the outer outline. Then with **Copy** make copies to the other corners.

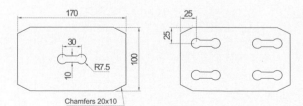

Fig. 5.52 Exercise 6

7. Construct the drawing shown in Fig. 5.53 and make two copies using **Copy**. With **Rotate** rotated each of the copies to the angles as shown.

8. Construct the dimensioned drawing of Fig. 5.54. With **Copy** copy the drawing. Then with **Scale** scale the drawing to a scale of **0.5**, followed by using **Rotate** to rotate the drawing through an angle of as shown. Finally, scale the original drawing to a scale of **2:1**.

9. Construct the left-hand drawing of Fig. 5.55. Include the dimensions in your drawing. Then, using the **Stretch** tool, stretch the drawing, including its dimensions to the sizes as shown in the right-hand drawing. The dimensions are said to be **associative** (see Chapter 6).

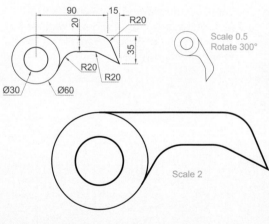

Fig. 5.54 Exercise 8

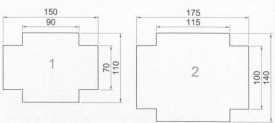

Fig. 5.55 Exercise 9

10. Construct the drawing Fig. 5.56. All parts of the drawing are plines of width = 0.7. The setting in the **Array** dialog is to be **180** in the **Angle of array** field.

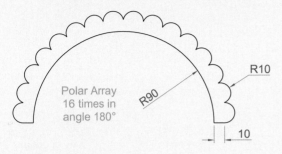

Fig. 5.56 Exercise 10

Chapter 6

Dimensions and Text

AIMS OF THIS CHAPTER

The aims of this chapter are:

1. to describe a variety of methods of dimensioning drawings;
2. to describe methods of adding text to drawings.

Introduction

The dimension style (**My_style**) has already been set in the acadiso.dwt template, which means that dimensions can be added to drawings using this dimension style.

The **Dimension tools**

There are several ways in which the dimensions tools can be called:

1. From the **Annotate/Dimensions** panel (Fig. 6.1).

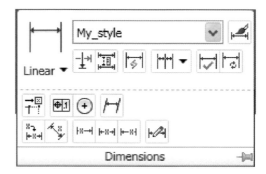

Fig. 6.1 Dimension tools in the **Annotate/Dimension** panel

2. From the **Dimension** toolbar (Fig. 6.2). The toolbar can be called to screen with a *right-click* in any toolbar on screen followed by a *click* on **Dimension** in the pop-up menu which appears.

Fig. 6.2 The **Dimension** toolbar

3. *Click* **Dimension** in the menu bar. Tools can be selected from the drop-down menu which appears.
4. By *entering* an abbreviation for a dimension tool at the command line.

Any one of these methods can be used when dimensioning a drawing, but some operators may well decide to use a combination of the four methods.

Adding dimensions using the tools

First example – **Linear Dimension** (Fig. 6.4)

1. Construct a rectangle 180 × 110 using the **Polyline** tool.
2. Make the **Dimensions** layer current (**Home/Layers** panel).

3. *Click* the **Linear** tool icon in the **Annotate/Dimension** panel (Fig. 6.3) or on **Linear** in the **Dimension** toolbar. The command line shows:

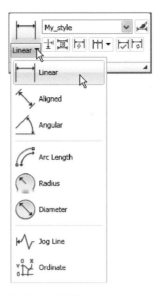

Fig. 6.3 The **Linear** tool icon in the **Annotate/Dimension** panel

```
Command: _dimlinear
Specify first extension line origin or <select
  object>: pick
Specify second extension line origin: pick
Non-associative dimension created.
Specify dimension line location or [Mtext/Text/
  Angle/Horizontal/Vertical/Rotated]: pick
Dimension text = 180
Command:
```

Figure 6.4 shows the 180 dimension. Follow exactly the same procedure for the 110 dimension.

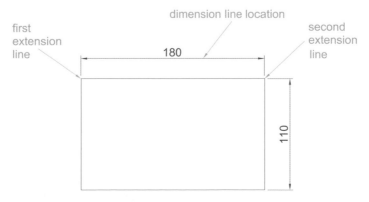

Fig. 6.4 First example – **Linear** dimension

> **Note**
>
> 1. If necessary use **Osnaps** to locate the extension line locations.
>
> 2. The prompt
>
>    ```
>    Specify first extension line origin or [select
>    object]:
>    ```
>
>   also allows the line being dimensioned to be picked.
>
> 3. The drop-down menu from the **Line** tool icon contains the following tool icons – **Angular**, **Linear**, **Aligned**, **Arc Length**, **Radius**, **Diameter**, **Jog Line** and **Ordinate**. Please refer to Fig. 6.3 when working through the examples below. **Note** when a tool is chosen from this menu, the icon in the panel changes to the selected tool icon.

Second example – **Aligned Dimension** (Fig. 6.5)

1. Construct the outline shown in Fig. 6.5 using the **Line** tool.
2. Make the **Dimensions** layer current (**Home/Layers** panel).
3. *Left-click* the **Aligned** tool icon (see Fig. 6.3) and dimension the outline. The prompts and replies are similar to the first example.

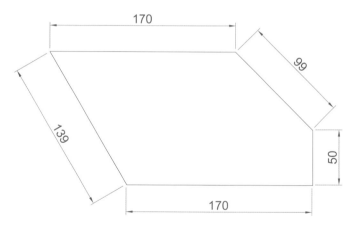

Fig. 6.5 Second example – **Aligned** dimension

Third example – **Radius Dimension** (Fig. 6.6)

1. Construct the outline in Fig. 6.8 using the **Line** and **Fillet** tools.
2. Make the **Dimensions** layer current (**Home/Layers** panel).
3. *Left-click* the **Radius** tool icon (see Fig. 6.3). The command line shows:

```
Command:_dimradius
Select arc or circle: pick one of the arcs
Dimension text = 30
```

```
Specify dimension line location or [Mtext/Text/
  Angle]: pick
Command:
```

4. Continue dimensioning the outline as shown in Fig. 6.6.

Fig. 6.6 Third example – **Radius** dimension

Note

1. At the prompt:

[Mtext/Text/Angle]:

If a **t** (Text) is *entered*, another number can be *entered*, but remember if the dimension is a radius the letter **R** must be *entered* as a prefix to the new number.

2. If the response is **a** (Angle), and an angle number is *entered* the text for the dimension will appear at an angle. Figure 6.7 show a radius dimension *entered* as an angle of 45°.

3. If the response is **m** (Mtext) the **Text Formatting** dialog appears together with a box in which new text can be *entered* (see p. 136).

4. Dimensions added to a drawing using other tools from the **Dimensions** control panel or from the **Dimension** toolbar should be practised.

Adding dimensions from the command line

From Figs. 6.1 and 6.2 it will be seen that there are some dimension tools which have not been described in examples. Some operators may prefer *entering* dimensions from the command line. This involves abbreviations for the required dimension such as:

- For **Linear Dimension** – **hor** (horizontal) or **ve** (vertical)
- For **Aligned Dimension** – **al**
- For **Radius Dimension** – **ra**

CHAPTER 6

- For **Diameter Dimension** – **d**
- For **Angular Dimension** – **an**
- For **Dimension Text Edit** – **te**
- For **Quick Leader** – **l**
- And **to** exit from the dimension commands – **e** (Exit).

First example – **hor** and **ve** (horizontal and vertical) – Fig. 6.8

1. Construct the outline Fig. 6.7 using the **Line** tool. Its dimensions are shown in Fig. 6.8.

Fig. 6.7 First example – outline to dimension

2. Make the **Dimensions** layer current (**Home/Layers** panel).
3. At the command line *enter* **dim**. The command line will show:

```
Command: enter dim right-click
Dim: enter hor (horizontal) right-click
Specify first extension line origin or <select
  object>: pick
Specify second extension line origin: pick
Non-associative dimension created.
Specify dimension line location or [Mtext/Text/
  Angle]: pick
Enter dimension text <50>: right-click
Dim: right-click
HORIZONTAL
Specify first extension line origin or <select
  object>: pick
Specify second extension line origin: pick
Non-associative dimension created.
Specify dimension line location or [Mtext/Text/
  Angle/Horizontal/Vertical/Rotated]: pick
Enter dimension text <140>: right-click
Dim: right-click
```

and the 50 and 140 horizontal dimensions are added to the outline.

4. Continue to add the right-hand 50 dimension. Then when the command line shows:

```
Dim: enter ve (vertical) right-click
Specify first extension line origin or <select
  object>: pick
Specify second extension line origin: pick
Specify dimension line location or [Mtext/Text/
  Angle/Horizontal/Vertical/Rotated]: pick
Dimension text <20>: right-click
Dim: right-click
VERTICAL
Specify first extension line origin or <select
  object>: pick
Specify second extension line origin: pick
Specify dimension line location or [Mtext/Text/
  Angle/Horizontal/Vertical/Rotated]: pick
Dimension text <100>: right-click
Dim: enter e (Exit) right-click
Command:
```

The result is shown in Fig. 6.8.

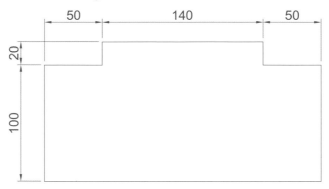

Fig. 6.8 First example – horizontal and vertical dimensions

Second example – an (Angular) – Fig. 6.10

1. Construct the outline Fig. 6.9 – a pline of width = 1.
2. Make the **Dimensions** layer current (**Home/Layers** panel).
3. At the command line:

```
Command: enter dim right-click
Dim: enter an right-click
Select arc, circle, line or <specify vertex>: pick
Select second line: pick
Specify dimension arc line location or [Mtext/
  Text/Angle/Quadrant]: pick
Enter dimension <90>: right-click
```

```
Enter text location (or press ENTER): pick
Dim:
```

and so on to add the other angular dimensions.

The result is given in Fig. 6.9.

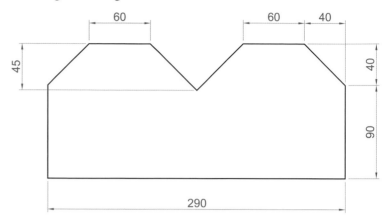

Fig. 6.9 Second example – outline for dimensions

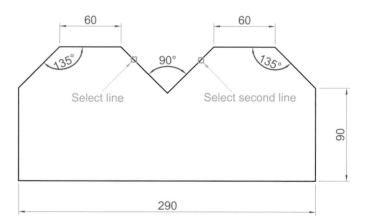

Fig. 6.10 Second example – **an** (Angle) dimension

Third example – l (Leader) – (Fig. 6.12)

1. Construct Fig. 6.11.
2. Make the **Dimensions** layer current (**Home/Layers** panel).
3. At the command line:

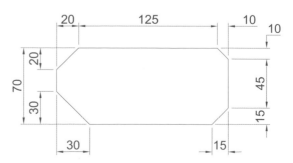

Fig. 6.11 Third example – outline for dimensioning

```
Command: enter dim right-click
Dim: enter l (Leader) right-click
Leader start: enter nea (osnap nearest) right-click
  to pick one of the chamfer lines
To point: pick
To point: pick
To point: right-click
Dimension text <0>: enter CHA 10 × 10
  right-click
Dim: right-click
```

Continue to add the other leader dimensions – Fig. 6.12.

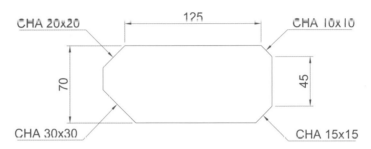

Fig. 6.12 Third example – **l** (Leader) dimensions

Fourth example – **te** (dimension text edit) – (Fig. 6.14)

1. Construct Fig. 6.13.

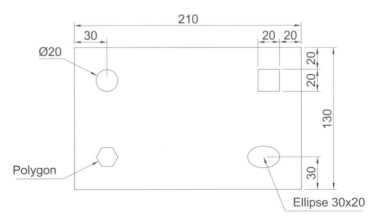

Fig. 6.13 Fourth example – dimensioned drawing

2. Make the **Dimensions** layer current (**Home/Layers** panel).
3. At the command line:

```
Command: enter dim right-click
Dim: enter te (tedit) right-click
Select dimension: pick the dimension to be changed
```

```
Specify new location for text or [Left/Right/
Center/Home/Angle]: either pick or enter a prompt
capital letter
Dim:
```

The results as given in Fig. 6.14 show dimensions which have been moved. The **210** dimension changed to the left-hand end of the dimension line, the **130** dimension changed to the left-hand end of the dimension line and the **30** dimension position changed.

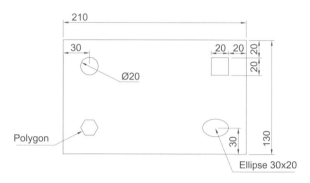

Fig. 6.14 Fourth example – dimensions amended with **tedit**

The **Arc Length** tool (Fig. 6.15)

1. Construct two arcs of different sizes as in Fig. 6.15.
2. Make the **Dimensions** layer current (**Home/Layers** panel).
3. Call the **Arc Length** tool from the **Annotate/Dimensions** panel (see Fig. 6.3), or *click* on **Arc Length** in the **Dimension** toolbar or *enter* **dimarc** at the command line. The command line shows:

```
Command: _dimarc
Select arc or polyline arc segment: pick an arc
Specify arc length dimension location, or [Mtext/
  Text/Angle/Partial/Leader]: pick a suitable
  position
Dimension text = 147
Command:
```

Examples on two arcs are shown in Fig. 6.15.

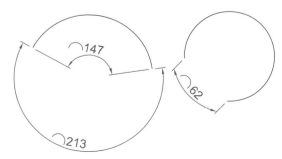

Fig. 6.15 Examples – **Arc Length** tool

The **Jogged** tool (Fig. 6.16)

1. Draw a circle and an arc as indicated in Fig. 6.16.
2. Make the **Dimensions** layer current (**Home/Layers** panel).
3. Call the **Jogged** tool, either with a *left-click* on its tool icon in the **Annotation/Dimension** panel see (Fig. 6.3) or with a *click* on **Jogged** in the **Dimension** toolbar or by *entering* **jog** at the command line. The command line shows:

```
Command: _dimjogged
Select arc or circle: pick the circle or the arc
Specify center location override: pick
Dimension text = 60
Specify dimension line location or [Mtext/Text/
  Angle]: pick
Specify jog location: pick
Command:
```

The results of placing as jogged dimension on a circle and an arc are shown in Fig. 6.16.

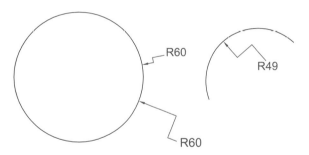

Fig. 6.16 Examples – the **Jogged** tool

Dimension tolerances

Before simple tolerances can be included with dimensions, new settings will need to be made in the **Dimension Style Manager** dialog as follows:

1. Open the dialog. The quickest way of doing this is to *enter* **d** at the command line followed by a *right-click*. This opens up the dialog.
2. *Click* the **Modify...** button of the dialog, followed by a *left-click* on the **Primary Units** tab and in the resulting sub-dialog make settings as shown in Fig. 6.17. Note the changes in the preview box of the dialog.

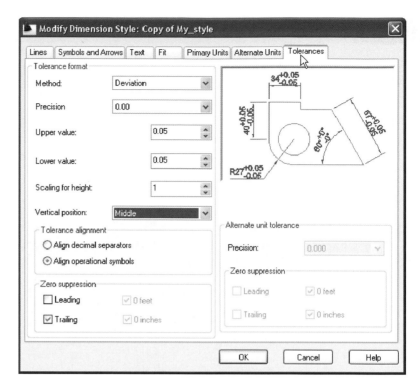

Fig. 6.17 The **Tolerances** subdialog of the **Modify Dimension Style** dialog

Example – tolerances (Fig. 6.19)

1. Construct the outline Fig. 6.18.

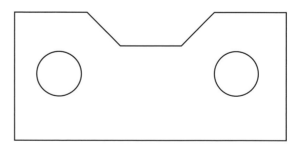

Fig. 6.18 First example – simple tolerances – outline

2. Make the **Dimensions** layer current (**Home/Layers** panel).
3. Dimension the drawing using either tools from the **Dimension** toolbar or by *entering* abbreviations at the command line. Because tolerances have been set in the **Dimension Style Manager** dialog, the toleranced dimensions will automatically be added to the drawing (Fig. 6.19).

Text

There are two main methods of adding text to drawings – **Multiline Text** and **Single Line Text**.

Example – **Single Line Text** (Fig. 6.23)

1. Open the drawing from the example on tolerances – Fig. 6.19.
2. Make the **Text** layer current (**Home/Layers** panel).
3. At the command line *enter* **dt** (for Single Line Text) followed by a *right-click*:

```
Command: enter dt right-click
TEXT
Current text style "ARIAL" Text height: 8
  Annotative No:
Specify start point of text or [Justify/Style]:
  pick
Specify rotation angle of text <0>: right-click
Enter text: enter The dimensions in this drawing
  show tolerances press the Return key twice
Command:
```

The result is given in Fig. 6.19.

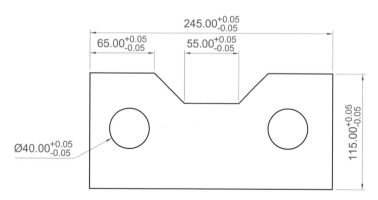

The dimensions in this drawing show tolerances

Fig. 6.19 Example – tolerances

Note

1. When using **Dynamic Text** the **Return** key of the keyboard is pressed when the text has been *entered*. A *right-click* does not work.

2. At the prompt:

```
Specify start point of text or [Justify/
  Style]: enter s (Style) right-click
```

`Enter style name or [?] <ARIAL>:` *`enter`* `Romand` *`right-click`*

`Enter text style(s) to list <*>:` *`right-click`*

an **AutoCAD Text Window** (Fig. 6.20) appears listing all the styles which have been selected in the **Text Style** dialog (see p. 135).

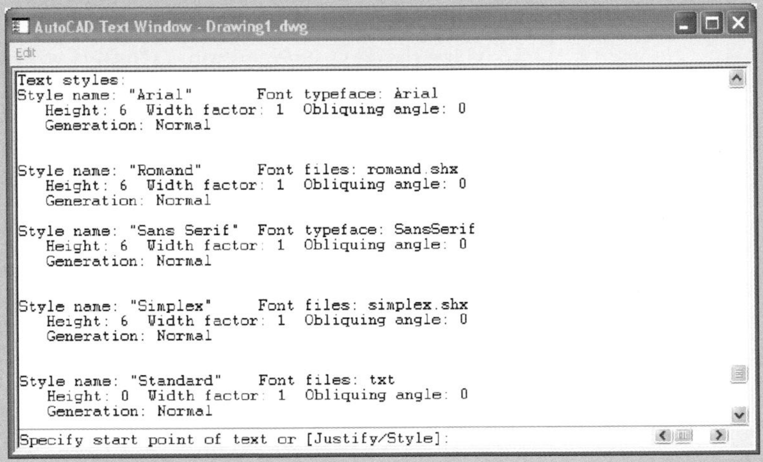
AutoCAD Text Window - Drawing1.dwg

Edit

```
Text styles:
Style name: "Arial"        Font typeface: Arial
   Height: 6  Width factor: 1  Obliquing angle: 0
   Generation: Normal

Style name: "Romand"       Font files: romand.shx
   Height: 6  Width factor: 1  Obliquing angle: 0
   Generation: Normal

Style name: "Sans Serif"  Font typeface: SansSerif
   Height: 6  Width factor: 1  Obliquing angle: 0
   Generation: Normal

Style name: "Simplex"      Font files: simplex.shx
   Height: 6  Width factor: 1  Obliquing angle: 0
   Generation: Normal

Style name: "Standard"     Font files: txt
   Height: 0  Width factor: 1  Obliquing angle: 0
   Generation: Normal
Specify start point of text or [Justify/Style]:
```

Fig. 6.20 The **AutoCAD Text Window**

3. In order to select the required text style its name must be *entered* at the prompt:

   `Enter style name or [?] <ARIAL>:` *`enter`* `Romand` *`right-click`*

And the text *entered* will be in the **Romand** style of height **9**, but only if that style were previously been selected in the **Text Style** dialog.

4. Figure 6.21 shows some text styles from the **AutoCAD Text Window**.

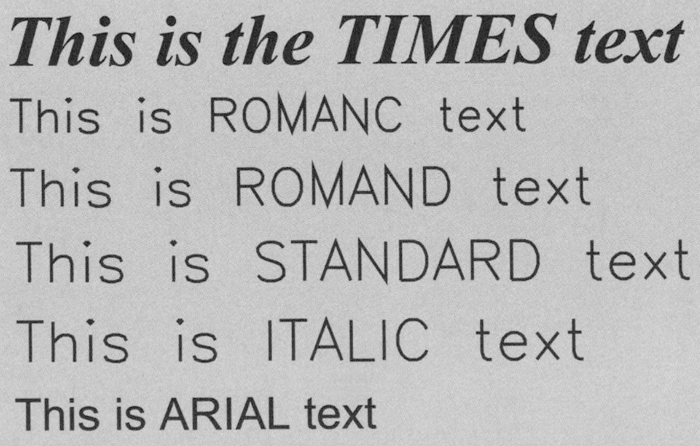
This is the TIMES text
This is ROMANC text
This is ROMAND text
This is STANDARD text
This is ITALIC text
This is ARIAL text

Fig. 6.21 Some text styles

5. There are two types of text fonts available in AutoCAD 2009 – the **AutoCAD SHX** fonts and the **Windows True Type** fonts.

The styles shown in Fig. 6.21, the **ITALIC**, **ROMAND**, **ROMANS** and **STANDARD** styles are AutoCAD text fonts. The **TIMES** and **ARIAL** styles are **Windows True Type** styles. Most of the **True Type** fonts can be *entered* in **Bold**, **Bold Italic**, **Italic** or **Regular** styles, but these variations are not possible with the AutoCAD fonts.

6. In the **Font name** pop-up list of the **Text Style** dialog, shows that a large number of text styles are available to the AutoCAD 2009 operator. It is advisable to practise using a variety of these fonts to familiarize oneself with the text opportunities available with AutoCAD 2009.

Example – Multiline Text (Fig. 6.23)

1. Make the **Text** layer current (**Home/Layers** panel).
2. Either *left-click* on the **Multiline Text** tool icon in the **Home/Annotation** panel (Fig. 6.22) or *click* on **Multiline Text…** in the **Draw** toolbar or *enter* **t** at the command line:

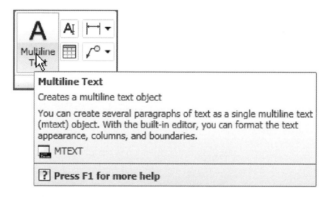

Fig. 6.22 Selecting **Multiline Text…** from the **Home/Annotation** panel

```
Command:_mtext
Current text style: "Arial" Text height:
  6 Annotative No
Specify first corner: pick
Specify opposite corner or [Height/Justify/Line
  spacing/Rotation/Style/Width/Columns]: pick
```

As soon as the **opposite corner** is *picked*, the **Text Formatting** box appears (Fig. 6.23). Text can now be *entered* as required within the box as indicated in this illustration.

When all the required text has been *entered*, *left-click* and the text box disappears, leaving the text on screen.

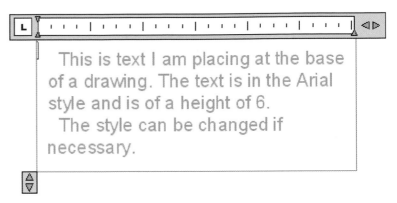

Fig. 6.23 Example – **Multiline Text** *entered* in the text box

Symbols used in text

When text has to be added by *entering* letters and figures as part of a dimension, the following symbols must be used:

- To obtain **Ø75** *enter* **%%c75**
- To obtain **55%** *enter* **55%%%**
- To obtain **±0.05** *enter* **%%p0.05**
- To obtain **90°** *enter* **60%%d**.

Checking spelling

Note

When a misspelt word or a word not in the AutoCAD spelling dictionary is *entered* in the **Multiline Text** box, red dots appear under the word, allowing immediate correction.

There are two methods for the checking of spelling in AutoCAD 2009.

First example – spell checking – **ddedit** (Fig. 6.24)

1. *Enter* some badly spelt text as indicated in Fig. 6.24.
2. *Enter* **ddedit** at the command line.
3. *Left-click* on the text. The text is highlighted. Edit the text as if working in a word-processing application and when satisfied *left-click* followed by a *right-click*.

Second example – the spelling tool (Fig. 6.24)

1. *Enter* some badly spelt text as indicated in Fig. 6.24.
2. Either *click* the **Spell Check...** icon in the **Annotate/Text** control (Fig. 6.26) or *enter* **spell** or **sp** at the command line.

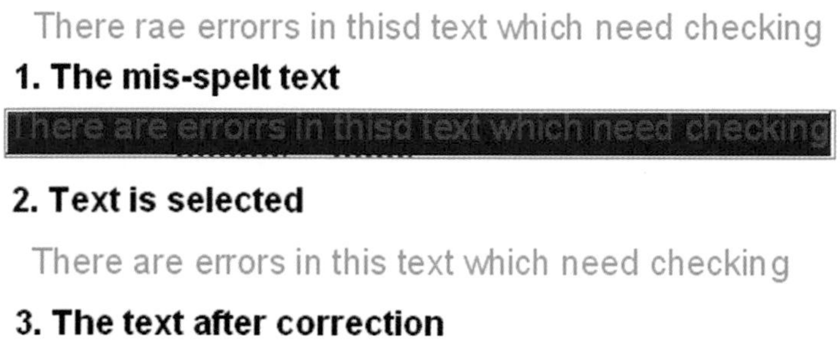

There rae errorrs in thisd text which need checking

1. The mis-spelt text

There are errorrs in thisd text which need checking

2. Text is selected

There are errors in this text which need checking

3. The text after correction

Fig. 6.24 First example – **spell checking** – **ddedit**

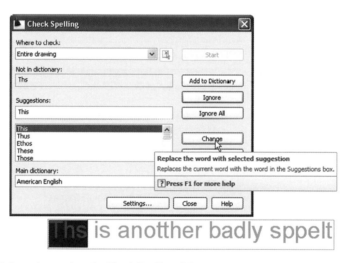

Fig. 6.25 Second example – the **Check Spelling** dialog

Fig. 6.26 The **Check Spelling** icon in the **Annotate/Text** panel

CHAPTER 6

AutoCAD Message ☒

Spelling check complete.

OK

Fig. 6.27 The **AutoCAD Message** window showing that spelling check is complete

3. The **Check Spelling** dialog appears (Fig. 6.25). In the **Where to look** field select **Entire drawing** from the field's pop-up list. The first badly spelt word is highlighted with words to replace them listed in the **Suggestions** field. Select the appropriate correct spelling as shown. Continue until all text is checked. When completely checked an **AutoCAD Message** appears (Fig. 6.27) If satisfied *click* its **OK** button.

REVISION NOTES

1. In the **Line and Arrows** sub-dialog of the **Dimension Style Manager** dialog **Lineweights** were set to **0.3** (p. 81). If these lineweights are to show in the drawing area of AutoCAD 2009, the **Show/Hide Lineweight** button in the status bar must be set **ON**.
2. Dimensions can be added to drawings using the tools from the **Annotate/Dimensions** panel, from the **Dimension** toolbar, or by *entering* **dim**, followed by abbreviations for the tools at the command line.
3. It is usually advisable to use osnaps when locating points on a drawing for dimensioning.
4. The **Style** and **Angle** of the text associated with dimensions can be changed during the dimensioning process.
5. When wishing to add tolerances to dimensions it will probably be necessary to make new settings in the **Dimension Style Manager** dialog.
6. There are two methods for adding text to a drawing – **Single Line Text** and **Multiline Text**.
7. When adding **Single Line Text** to a drawing, the **Return** key must be used and not the right-hand mouse button.
8. Text styles can be changed during the process of adding text to drawings.
9. AutoCAD 2009 uses two types of text style – **AutoCAD SHX** fonts and **Windows True Type** fonts.
10. Most **True Type** fonts can be in **bold**, ***bold italic,*** *italic* or regular format. **AutoCAD** fonts can only be added in the single format.
11. To obtain the symbols **Ø**; **±**; °; **%** use **%%c**; **%%p**; **%%d**; **%%%** before the figures of the dimension.
12 Text spelling can be checked with by selecting **Text/Edit...** from the **Modify** drop-down menu, by selecting **Spell Check...** from the **Text** control panel, or by *entering* **spell** or **sp** at the command line.

Exercises

Methods of constructing answers to the following exercises can be found in the free website:

http://books.elsevier.com/companions/9780750689830

1. Open any of the drawings previously saved from working through examples or as answers to exercises and add appropriate dimensions.

2. Construct the drawing shown in Fig. 6.28 but in place of the given dimensions add dimensions showing tolerances of 0.25 above and below.

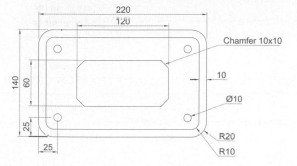

Fig. 6.28 Exercise 2

3. Construct and dimension the drawing of Fig. 6.29.

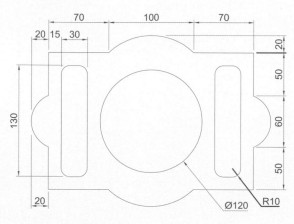

Fig. 6.29 Exercise 3

4. Construct two polygons as in Fig. 6.30 and add all diagonals. Set osnaps **endpoint**

and **intersection** and using the lines as in Fig. 6.30 construct the stars as shown using a polyline of width = 3. Next erase

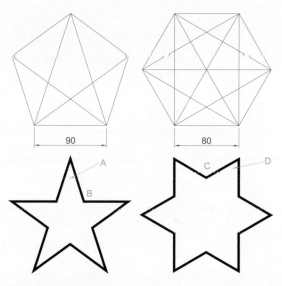

Fig. 6.30 Exercise 4

all unwanted lines. Dimension the angles labelled **A**, **B**, **C** and **D**.

5. Using the text style **Arial** of height **20** and enclosing the wording within a pline rectangle of Width = 5 and Fillet = 10, construct the drawing shown in Fig. 6.31.

Fig. 6.31 Exercise 5

Chapter 7

Orthographic and isometric

AIM OF THIS CHAPTER

The aims of this chapter are to introduce methods of constructing views in orthographic projection and the construction of isometric drawings.

Orthographic projection

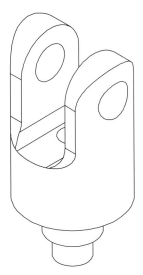

Fig. 7.1 Example –
orthographic projection –
the solid being drawn

Orthographic projection involves viewing an article being described in a technical drawing from different directions – from the front, from a side, from above, from below or from any other viewing position. Orthographic projection often involves:

- the drawing of details which are hidden, using hidden detail lines
- sectional views in which the article being drawn is imagined as being cut through and the cut surface drawn
- centre lines through arcs, circles spheres and cylindrical shapes.

An example of an orthographic projection

Taking the solid shown in Fig. 7.1, to construct a three-view orthographic projection of the solid:

1. Draw what is seen when the solid is viewed from its left-hand side and regard this as the **front** of the solid. What is drawn will be a **front view** (Fig. 7.2).

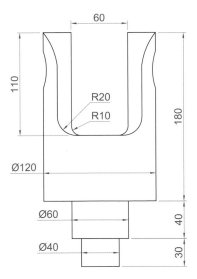

Fig. 7.2 The **front view** of the solid

2. Draw what is seen when the solid is viewed from the left-hand end of the front view. This produces an **end view**. Figure 7.3 shows the end view alongside the front view.
3. Draw what is seen when the solid is viewed from above the front view. This produces a **plan**. Figure 7.4 shows the plan below the front view.
4. In the **Home/Layers** panel, in the **Layer Control** list *click* on **Center** to make it the current layer (Fig. 7.5). All lines will now be drawn as centre lines.
5. In the three-view drawing add centre lines.
6. Make the **Hidden** layer the current layer and add hidden detail lines.

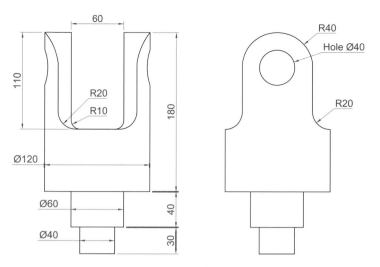

Fig. 7.3 **Front** and **end** views of the solid

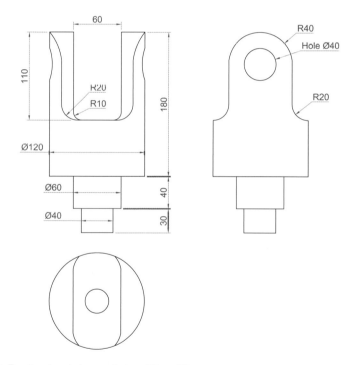

Fig. 7.4 **Front** and **end** views and **plan** of the solid

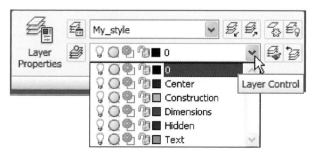

Fig. 7.5 Making the layer **Center** current from the **Home/Layers** panel

7. Make the **Text** layer current and add border lines and a title block.
8. Make the **Dimensions** layer current and add all dimensions.

The completed drawing is shown in Fig. 7.6.

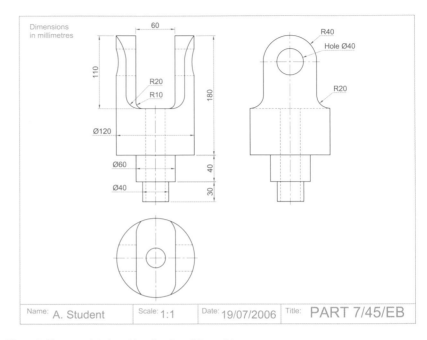

Fig. 7.6 The completed working drawing of the solid

First angle and third angle

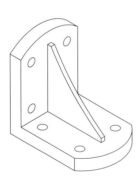

Fig. 7.7 The solid used to demonstrate first and third angles of projection

There are two types of orthographic projection – **first angle** and **third angle**. Figure 7.7 is a pictorial drawing of the solid used to demonstrate the two angles. Figure 7.8 shows a three-view **first angle projection** and Fig. 7.9 the same views in **third angle**.

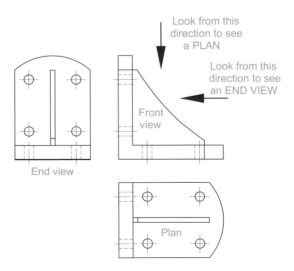

Fig. 7.8 A **first angle** projection

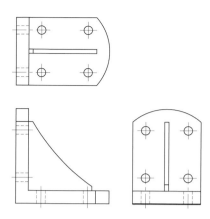

Fig. 7.9 A **third angle** projection

In both angles the viewing is from the same directions. The difference is that the view as seen is placed on the viewing side of the front view in **third angle** and on the opposite side to the viewing in **first angle**.

Sectional views

In order to show internal shapes of a solid being drawn in orthographic projection the solid is imagined as being cut along a plane and the cut surface then drawn as seen. Common practice is to **hatch** the areas which then show in the cut surface. Note the section plane line, the section label and the hatching in the sectional view (Fig. 7.10).

Adding hatching

To add the hatching as shown in Fig. 7.10:

1. Call the **Hatch** tool – either *left-click* on its tool icon in the **Home/ Draw** panel (Fig. 7.11), *click* the tool in the **Draw** toolbar, or *enter* **h**

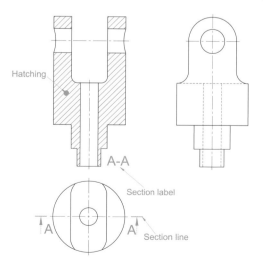

Fig. 7.10 A sectional view

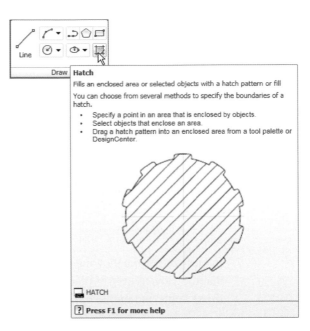

Fig. 7.11 The **Hatch** tool icon and tooltip from the **Home/Draw** panel

at the command line. **Note:** do not *enter* **hatch** as this gives a different result. The **Hatch and Gradient** dialog (Fig. 7.12) appears.

2. *Click* in the **Swatch** field. The **Hatch Pattern Palette** appears. *Left-click* the **ANSI** tab and from the resulting pattern icons *double-click* the **ANSI31** icon. The palette disappears and the **ANSI31** pattern appears in the **Swatch** field.

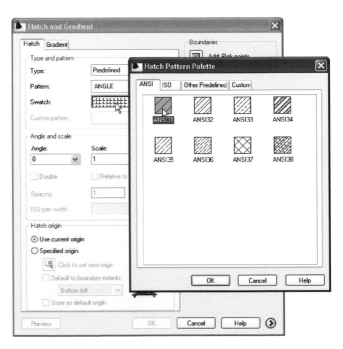

Fig. 7.12 The **Hatch and Gradient** dialog and the **ANSI Hatch Pattern Palette**

3. In the dialog *left-click* the **Pick an internal point** button (Fig. 7.13). The dialog disappears.

4. In the front view *pick* points as shown in the left-hand drawing of Fig. 7.14. The dialog reappears. *Click* the **Preview** button of the dialog and in the sectional view which reappears, check whether the hatching is satisfactory. In this example it may well be that the **Scale** figure in the dialog needs to be *entered* as **2** in place of the default **1**. Press the **Esc** key of the keyboard and the dialog returns. Change the figure and **Preview** again. If satisfied *right-click*.

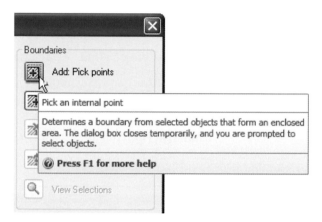

Fig. 7.13 The **Pick an internal point** button of the **Boundary Hatch and Fill** dialog

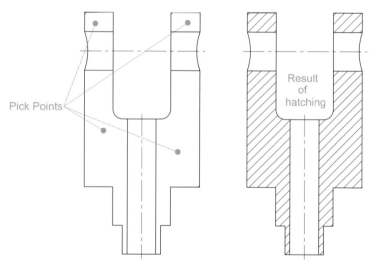

Fig. 7.14 The result of hatching

Isometric drawing

Isometric drawing must not be confused with solid model drawing, examples of which are given in Chapters 12 to 19. Isometric drawing is a 2D method of describing objects in a pictorial form.

Setting the AutoCAD window for isometric drawing

To set the AutoCAD 2009 window for the construction of isometric drawings:

1. At the command line:

```
Command: enter snap
Specify snap spacing or [On/Off/Aspect/Rotate/
  Style/Type] <5>: s (Style)
Enter snap grid style [Standard/Isometric]
  <S>: i (Isometric)
Specify vertical spacing <5>: right-click
Command:
```

and the grid dots in the window assume an isometric pattern as shown in Fig. 7.15. Note also the cursor hair lines which are at set in an **Isometric right** angle.

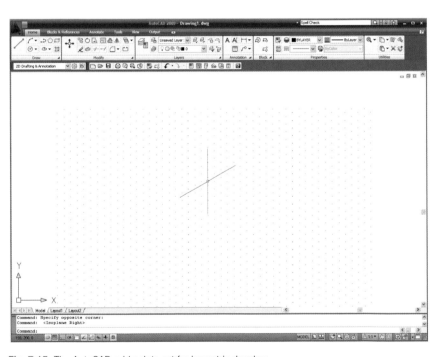

Fig. 7.15 The AutoCAD grid points set for isometric drawing

Fig. 7.16 The three isoplanes

2. There are three isometric angles – **Isoplane Top**, **Isoplane Left** and **Isoplane Right**. These can be set either by pressing the **F5** function key or by pressing the **Ctrl** and **E** keys. Repeated pressing of either of these 'toggles' between the three settings. Figure 7.16 is an isometric view showing the three isometric planes.

The isometric circle

Circles in an isometric drawing show as ellipses. To add an isometric circle to an isometric drawing, call the **Ellipse** tool. The command line shows:

```
Command: _ellipse
Specify axis endpoint of ellipse or [Arc/Center/
  Isocircle]:enter i (Isocircle) right-click
Specify center of isocircle:pick or enter
  coordinates
Specify radius of isocircle or [Diameter]: enter
  a number
Command:
```

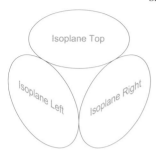

Fig 7 17 The three isocircles

and the isocircle appears. Its isoplane position is determined by which of the isoplanes is in operation at the time the isocircle was formed. Figure 7.17 shows these three isoplanes containing isocircles.

Examples of isometric drawings

First example – isometric drawing (Fig. 7.20)

1. Working to the shapes and sizes given in the orthographic projection in Fig. 7.18, set **Snap** on (press the **F9** function key) and **Grid** on (**F7**).
2. Set **Snap** to **Isometric** and set the isoplane to **Isoplane Top** using **F5**.
3. With **Line**, construct the outline of the top of the model (Fig. 7.19) working to the dimensions given in Fig. 7.18.

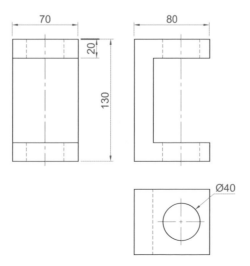

Fig. 7.18 First example – isometric drawing – the model

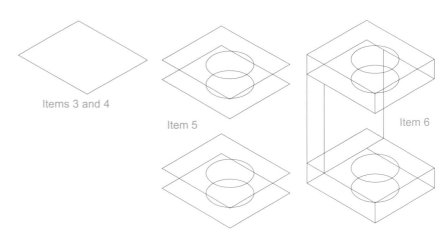

Items 3 and 4

Item 5

Item 6

Fig. 7.19 First example – isometric drawing – items 3, 4, 5 and 6

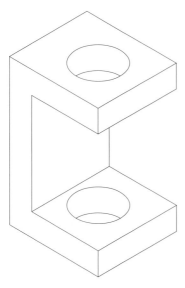

Fig. 7.20 First example – isometric drawing

4. Call **Ellipse** tool and set to isocircle and add the isocircle of radius 20 centred in its correct position in the outline of the top (Fig. 7.29).
5. Set the isoplane to **Isoplane Right** and with the **Copy** tool, copy the top with its ellipse vertically downwards three times as shown in Fig. 7.20.
6. Add lines as shown in Fig. 7.19.
7. Finally using **Trim** remove unwanted parts of lines and ellipses to produce Fig. 7.20.

Second example – isometric drawing (Fig. 7.22)

Figure 7.21 is an orthographic projection of the model of which the isometric drawing is to be constructed. Figure 7.22 shows the stages in its construction. The numbers refer to the items in the list below.

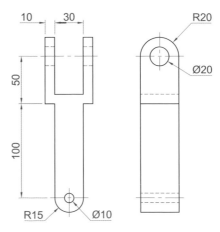

Fig. 7.21 Second example – isometric drawing – orthographic projection of model

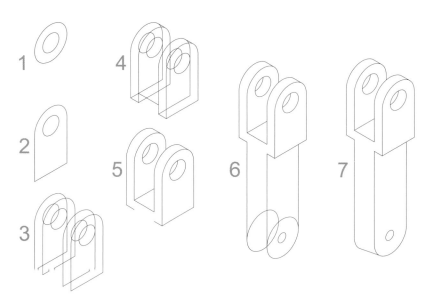

Fig. 7.22 Second example – isometric drawing – stages in the construction

1. In **Isoplane Right** construct two isocircles of radii 10 and 20.
2. Add lines as in drawing **2** and trim unwanted parts of isocircle.
3. With **Copy** copy 3 times as in drawing **3**.
4. With **Trim** trim unwanted lines and parts of isocircle as in drawing **4**.
5. In **Isoplane Left** add lines as in drawing **5**.
6. In **Isoplane Right** add lines and isocircles as in drawing **6.**
7. With **Trim** trim unwanted lines and parts of isocircles to complete the isometric drawing – drawing **7**.

REVISION NOTES

1. There are, in the main, two types of orthographic projection – first angle and third angle.
2. The number of views included in an orthographic projection depends upon the complexity of the component being drawn – a good rule to follow is to attempt to fully describe the object in as few views as possible.
3. Sectional views allow parts of an object which are normally hidden from view to be more fully described in a projection.
4. When a layer is turned **OFF**, all constructions on that layer disappear from the screen.
5. If a layer is locked, objects can be added to the layer but no further additions or modifications can be added to the layer. If an attempt is made to modify an object on a locked layer, a small lock icon appears near the object and the command line shows:

```
Command: _erase
Select objects:pick 1 found
1 was on a locked layer
```

and the object will not be modified.

6. Frozen layers cannot be selected, but note that layer **0** cannot be frozen.
7. Isometric drawing is a 2D pictorial method of producing illustrations showing objects. It is not a 3D method of showing a pictorial view.
8. When drawing ellipses in an isometric drawing the **Isocircle** prompt of the **Ellipse** tool command line sequence must be used.
9. When constructing an isometric drawing **Snap** must be set to **Isometric** mode before construction can commence.

Exercises

Methods of constructing answers to the following exercises can be found in the free website:

http://books.elsevier.com/companions/9780750689830

Figure 7.23 is an isometric drawing of a slider fitment on which the three exercises **1**, **2** and **3** are based

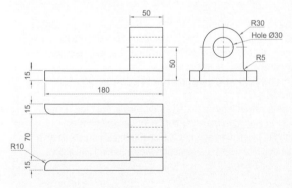

Fig. 7.23 Exercises 1, 2 and 3 – an isometric drawing of the three parts of the slider on which these exercises are based

1. Figure 7.24 shows a first angle orthographic projection of part of the fitment shown in the isometric drawing in Fig. 7.23. Construct a three-view third angle orthographic projection of the part.

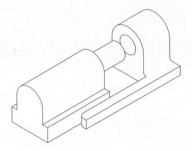

Fig. 7.24 Exercise 1

2. Figure 7.25 shows a first angle orthographic projection of the other part of the fitment. Construct a three-view third angle orthographic projection of the part.

3. Construct an isometric drawing of the part shown in Fig. 7.25.

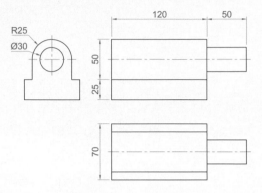

Fig. 7.25 Exercises 2 and 3

4. Construct a three-view orthographic projection in an angle of your own choice of the tool holder assembled as shown in the isometric drawing in Fig. 7.26. Details are given in Fig. 7.27.

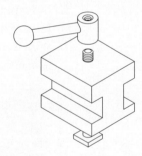

Fig. 7.26 Exercises 4 and 5 – orthographic projections of the three parts of the tool holder

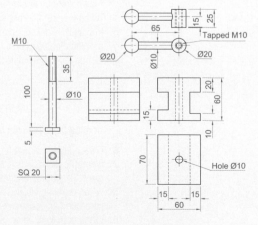

Fig. 7.27 Exercises 4 and 5 – orthographic drawing of the tool holder on which the two exercises are based

5. Construct an isometric drawing of the body of the tool holder shown in Figures 7.26 and 7.27.

6. Construct the orthographic projection given in Fig. 7.29.

7. Construct an isometric drawing of the angle plate shown in Figs 7.28 and 7.29.

8. Construct a third angle projection of the component shown in the isometric drawing in Fig. 7.30 and the three-view first angle projection shown in Fig. 7.31.

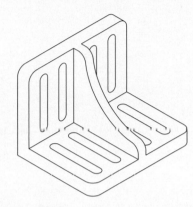

Fig. 7.28 An isometric drawing of the angle plate on which exercises 6 and 7 are based

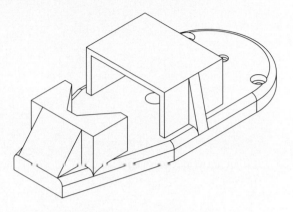

Fig. 7.30 Exercises 8 and 9 – an isometric drawing of the component for the two exercises

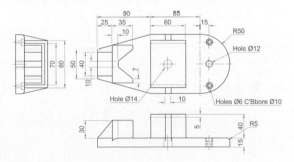

Fig. 7.31 Exercises 8 and 9

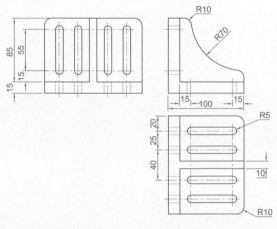

Fig. 7.29 Exercises 6 and 7 – an orthographic projection of the angle plate

9. Construct the isometric drawing shown in Fig. 7.30, working to the dimensions given in Fig. 7.31.

Chapter 8

Hatching

AIM OF THIS CHAPTER

The aim of this chapter is to give further examples of the use of hatching in its various forms.

Introduction

In Chapter 7 an example of the hatching of a sectional view in an orthographic projection was given. Further examples of the use of hatching will be described in this chapter.

There are a large number of hatch patterns available when hatching drawings in AutoCAD 2009. Some examples from the **Other Predefined** set of hatch patterns (Fig. 8.6) in the **Hatch Pattern Palette** sub-dialog are shown in Fig. 8.1.

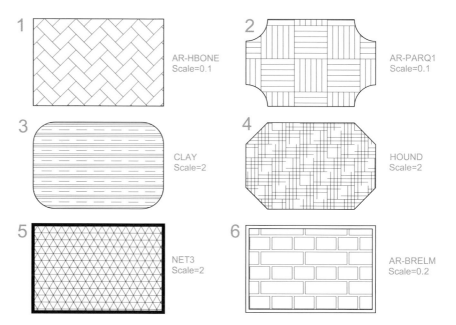

Fig. 8.1 Some hatch patterns from **Predefined** hatch patterns

Other hatch patterns can be selected from the **ISO** or **ANSI** hatch pattern palettes, or the operator can design his/her own hatch patterns and save them to the **Custom** hatch palette.

First Example – hatching a sectional view (Fig. 8.3)

Figure 8.3 shows a two-view orthographic projection which includes a sectional end view. Note the following in the drawing:

1. The section plane line, consisting of a centre line with its ends marked **A** and arrows showing the direction of viewing to obtain the sectional view.
2. The sectional view labelled with the letters of the section plane line.
3. The cut surfaces of the sectional view hatched with the **ANSI31** hatch pattern, which is in general use for the hatching of engineering drawing sections.

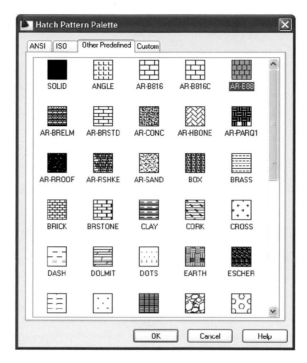

Fig. 8.2 The **Other Predefined** Hatch Pattern Palette

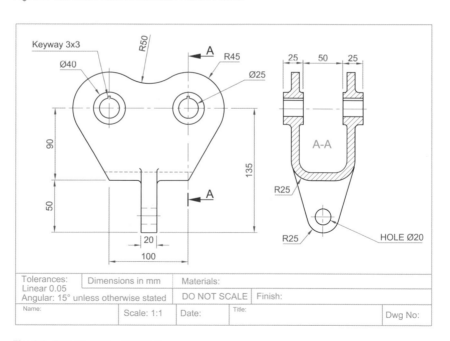

Fig. 8.3 First example – **Hatching**

Second example – hatching rules (Fig. 8.4)

Figure 8.4 describes the stages in hatching a sectional end view of a lathe tool holder. Note the following in the section:

1. There are two angles of hatching to differentiate in separate parts of the section.

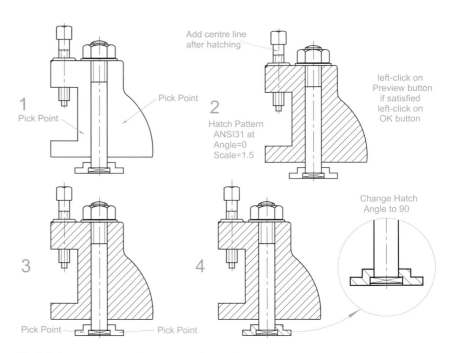

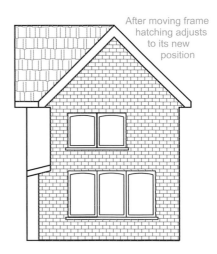

Fig. 8.4 Second example – hatching rules for sections

2. The section follows the general rule that parts such as screws, bolts, nuts, rivets, other cylindrical objects, webs and ribs and other such features are shown as outside views within sections.

Third example – Associative hatching (Fig. 8.5)

Figure 8.5 shows two end views of a house. After constructing the left-hand view, it was found that the upper window had been placed in the wrong position. Using the **Move** tool, the window was moved to a new position. The brick hatching automatically adjusted to the new position. Such **associative hatching** is only possible if check box is **ON** – a tick in the check box in the **Options** area of the dialog (Fig. 8.6).

Fig. 8.5 Third example – **Associative** hatching

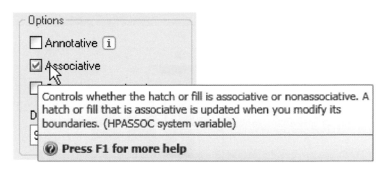

Fig. 8.6 **Associative** hatching set on in the **Hatch and Gradient** dialog

Fourth example – Colour gradient hatching (Fig. 8.9)

Fig. 8.8 shows two examples of hatching from the **Gradient** sub dialog of the **Hatch and Gradient** dialog.

1. Construct two outlines, each consisting of six rectangles (Fig. 8.9).
2. *Click* the **Gradient** icon in the **Home/Draw** panel (Fig. 8.7) or in the **Draw** toolbar. In the **Hatch and Gradient** dialog which appears (Fig. 8.8) *pick* one of the gradient choices, followed with a *click* on the **Pick an internal point** button. *Click* one of the color panels in the dialog and when the dialog disappears, *pick* a single area of one of the rectangles in the left-hand drawings, followed by a *click* on the dialog's **OK** button when the dialog reappears.
3. Repeat in each of the other rectangles of the left-hand drawing changing the pattern in each of the rectangles.

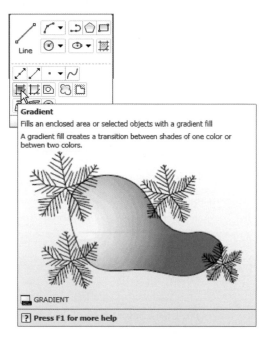

Fig. 8.7 The **Gradient...** tool icon from the **Home/Draw** panel

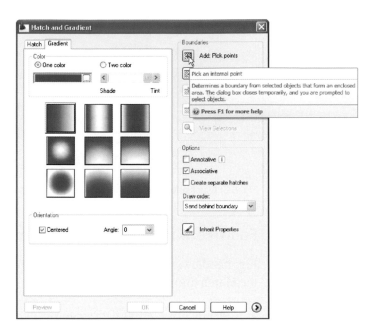

Fig. 8.8 The **Hatch and Gradient** dialog

4. *Click* the button (**...**) to the right of the **Color** field, select a new colour from the **Select Color** dialog which appears and repeat items **3** and **4** in six rectangles.

The result is shown in Fig. 8.9.

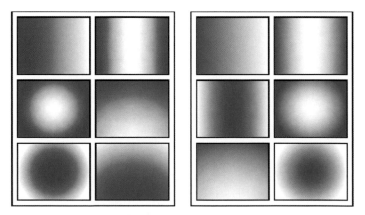

Fig. 8.9 Fourth example – **Gradient** hatching

> ### Note
>
> If the **Two color** radio button is set on (dot in circle) the colours involved in the gradient hatch can be changed by *clicking* the button marked with three full points (...) on the right of the colour field. This brings a **Select Color** dialog on screen, which offers three choices of sub-dialogs from which to select colours.

Fifth example – Advanced hatching (Fig. 8.12)

If the arrow at the bottom-right-hand corner of the **Hatch and Gradient** dialog is *clicked* (Fig. 8.10) the dialog expands to show the **Island display style** selections (Fig. 8.11).

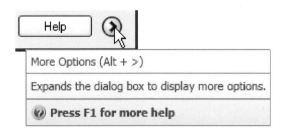

Fig. 8.10 The **More Options** arrow of the **Hatch and Gradient** dialog

Fig. 8.11 The **Island display style** selections in the expanded **Hatch and Gradient** dialog

1. Construct a drawing which includes three outlines as shown in the left-hand drawing of Fig. 8.12 and copy it twice to produce three identical drawings.
2. Select the hatch patterns **STARS** at an angle of **0** and scale **1**.
3. *Click* in the **Normal** radio button of the **Island display style** area.
4. *Pick* a point in the left-hand drawing. The drawing hatches as shown.
5. Repeat in the centre drawing with the radio button of the **Outer** style set on (dot in button).
6. Repeat in the right-hand drawing with **Ignore** set on.

Sixth example – Text in hatching (Fig. 8.13)

1. Construct a pline rectangle using the sizes given in Fig. 8.13.
2. In the **Text Style Manager** dialog, set the text font to **Arial** and its **Height = 25**.
3. Using the **Dtext** tool *enter* the text as shown central to the rectangle.
4. Hatch the area using the **HONEY** hatch pattern set to an angle of **0** and scale of **1**.

The result is shown in Fig. 8.13.

> **Note**
>
> Text will be entered with a surrounding boundary area free from hatching providing the **Normal** radio button of the **Island display style** selection is on.

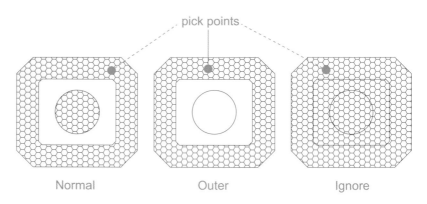

Fig. 8.12 Fifth example – **Advanced** hatching

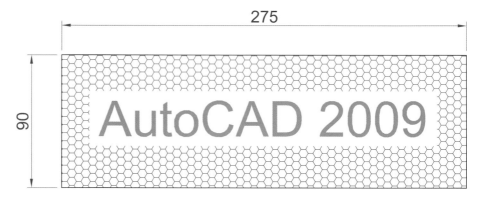

Fig. 8.13 Sixth example – Text in hatching

Seventh example – advanced hatching (Fig. 8.20)

1. From the **Home/Layers** panel open the **Layer** list with a *click* on the arrow to the right of the **Layer Control** field (Fig. 8.14).
2. Note the extra added layer **Hatch** colour red (Fig. 8.14).
3. With the layer **0** current construct the outline as given in Fig. 8.15.
4. Make layer **Text** current and construct the lines as shown in Fig. 8.16.

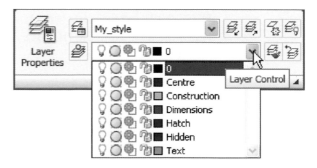

Fig. 8.14 Seventh example – the layers setup for the advanced hatch example

Fig. 8.15 Seventh example – construction on layer **0**

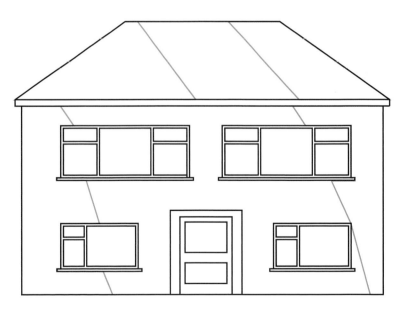

Fig. 8.16 Seventh example – construction on layer **Text**

5. Make the layer **Hatch** current and add hatching to the areas shown in Fig. 8.17 using the hatch patterns **ANGLE** at scale **2** for the roof and **BRICK** at a scale of **0.75** for the wall.

6. Finally turn the layer **Text** off. The result is given in Fig. 8.18.

Fig. 8.17 Seventh example – construction on layer **HATCH**

Fig. 8.18 Seventh example – the finished drawing

REVISION NOTES

1. A large variety of hatch patterns are available when working with AutoCAD 2009.
2. In sectional views in engineering drawings it is usual to show items such as bolts, screws, other cylindrical objects, webs and ribs as outside views.
3. When **Associative hatching** is set on, if an object is moved within a hatched area, the hatching accommodates to fit around the moved object.
4. **Colour gradient hatching** is available in AutoCAD 2009.
5. When hatching takes place around text, a space around the text will be free from hatching.

Exercises

Methods of constructing answers to the following exercises can be found in the free website:

http://books.elsevier.com/companions/9780750689830

1. Figure 8.19 shows a pictorial drawing of the component shown in the three-view orthographic projection of Fig. 8.20. Construct the three views but with the font view as a sectional view based on the section plane **A–A**.

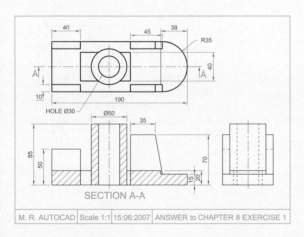

Fig. 8.19 Exercise – a pictorial view

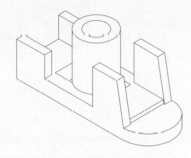

Fig. 8.20 Exercise 1

2. Construct the three-view orthographic projection shown in Fig. 8.21 to the given dimensions with the front view as the sectional view **A–A**.

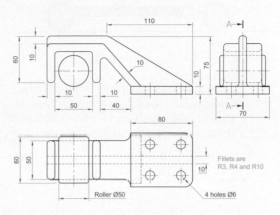

Fig. 8.21 Exercise 2

3. Construct the drawing **Stage 5** following the descriptions of stages given in Fig. 8.22.

Stage 1
Construct word on Layer 0 and offset on Layer 1

Stage 2
Hatch on Layer HATCH01 with SOLID Turn Layer 0 off

Stage 3
Turn HATCH01 off Turn Layer HATCH02 on Add lines as shown

Stage 4
On HATCH03 Hatch with ANSI31 at Angle 135 and Scale 40 Turn HATCH02 off

Stage 5
Turn HATCH02 on

Fig. 8.22 Exercise 3

4. Figure 8.23 is a front view of a car with parts hatched. Construct a similar drawing of any make of car, using hatching to emphasize the shape.

Fig. 8.23 Exercise 4

5. Working to the notes given with the drawing shown in Fig. 8.24, construct the end view of a house as shown. Use your own discretion about sizes for the parts of the drawing.

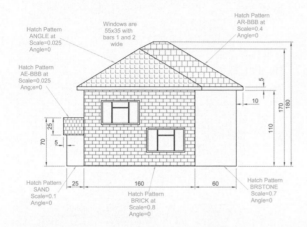

Fig. 8.24 Exercise 5

6. Working to dimensions of your own choice, construct the three-view projection of a two-storey house as shown in Fig. 8.25.

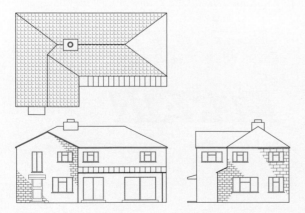

Fig. 8.25 Exercise 6

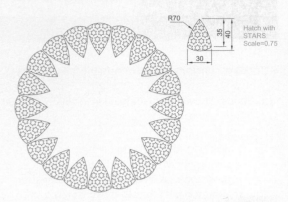

Fig. 8.26 Exercise 7

7. Construct Fig. 8.22 as follows:

    (a) On layer **Text**, construct a circle of radius **90**.
    (b) Make layer **0** current.
    (c) Construct the small drawing to the details as shown and save as a block with a block name **shape** (see Chapter 9).
    (d) Call the **Divide** tool by *entering* **div** at the command line:

```
Command: enter div right-click
Select object to divide: pick
  the circle
Enter number of segments or
  [Block]: enter b right-click
Enter name of block to insert:
  enter shape right-click
Align block with object? [Yes/
  No] <Y> : right-click
Enter the number of segments:
  enter 20 right-click
Command
```

    (e) Turn the layer **Text** off.

Blocks and Inserts

AIMS OF THIS CHAPTER

The aims of this chapter are:

1. to describe the construction of **blocks** and **wblocks** (written blocks);
2. to introduce the insertion of blocks and wblocks into drawings;
3. to introduce uses of the **DesignCenter** palette;
4. to explain the use of the **Explode** and **Purge** tools.

CHAPTER 9

Introduction

Blocks are drawings which can be inserted into other drawings. Blocks are contained in the data of the drawing in which they have been constructed. Wblocks (written blocks) are saved as drawings in their own right, but can be inserted into other drawings if required.

Blocks

First example – Blocks (Fig. 9.3)

1. Construct the building symbols as shown in Fig. 9.1 to a scale of 1:50.

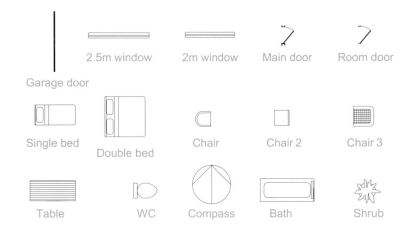

Fig. 9.1 First example – **Blocks** – symbols to be saved as blocks

2. *Left-click* the **Create** tool icon in the **Home/Block** panel (Fig. 9.2). The **Block Definition** dialog (Fig. 9.3) appears. To make a block from the **Double bed** symbol drawing:

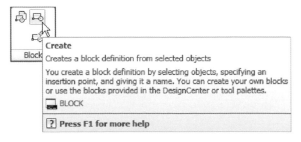

Fig. 9.2 Click **Create** tool icon in the **Home/Block** panel

(a) *Enter* **double bed** in the **Name** field.
(b) *Click* the **Select Objects** button. The dialog disappears. Window the drawing of the double bed. The dialog reappears. Note the icon of the double bed at the top-centre of the dialog.

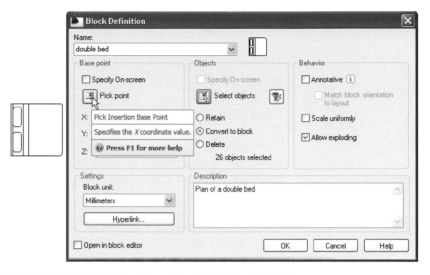

Fig. 9.3 The **Block Definiton** dialog with entries for the **double bed**

(c) *Click* the **Pick Point** button. The dialog disappears. *Click* a point on the double bed drawing to determine its **insertion point**. The dialog reappears.

(d) If thought necessary *enter* a description in the **Description** field of the dialog.

(e) *Click* the **OK** button. The drawing is now saved as a **block** in the drawing.

3. Repeat items **1** and **2** to make blocks of all the other symbols in the drawing.

4. Open the **Block Definition** dialog again and *click* the arrow on the right of the **Name** field. Blocks saved in the drawing are listed in Fig. 9.4.

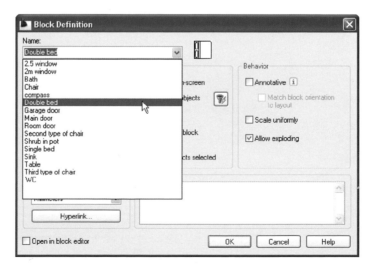

Fig. 9.4 The popup list in the **Name** field of the **Block Definition** dialog showing all blocks saved in the drawing

Inserting blocks into a drawing

There are two methods by which symbols saved as blocks can be inserted into another drawing.

Example – first method of inserting blocks

Ensure that all the symbols saved as blocks using the **Create** tool are saved in the data of the drawing in which the symbols were constructed, erase the drawings of the symbols and in their place construct the outline of the plan of a bungalow to a scale of 1:50 (Fig. 9.5). Then proceed as follows:

1. *Left-click* the **Insert** tool icon in the **Home/Block** panel (Fig. 9.6) or the **Insert Block** tool in the **Draw** toolbar. The **Insert** dialog appears on screen (Fig. 9.7). From the **Name** pop-up list select the name of the block which is to be inserted, in this example the **2.5 window**
2. Make sure the check box against **Explode** is off (no tick in box). *Click* the dialog's **OK** button the dialog disappears. The symbol drawing appears on screen with its insertion point at the intersection of the cursor hairs ready to be *dragged* into its position in the plan drawing
3. Once all the block drawings are placed, their positions can be adjusted. Blocks are single objects and can thus be dragged into new positions as

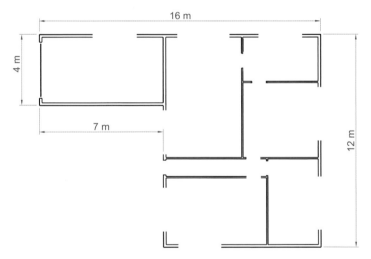

Fig. 9.5 First example – inserting blocks. Outline plan

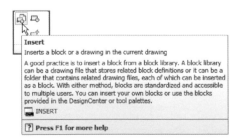

Fig. 9.6 The **Insert** tool icon in the **Home/Block** panel

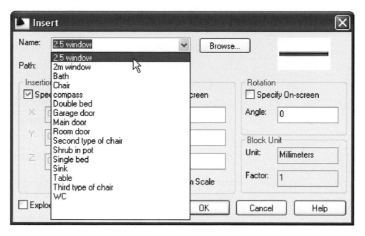

Fig. 9.7 The **Insert** dialog with its **Name** popup list displaying the names of all blocks in the
drawing

required under mouse control. Their angle of position can be amended
at the command line, which shows:

```
Command:
INSERT
Specify insertion point or [Basepoint/Scale/X/Y/Z/
  Rotate/PScale/PX/PY/PZ/PRotate]: enter r
  (Rotate) right-click
Specify insertion angle: enter 180 right-click
Specify insertion point: pick
Command:
```

Selection from these prompts allows scaling, stretching along any axis,
previewing etc., as the block is inserted.

4. Insert all necessary blocks and add other detail as required to the plan
outline drawing. The result is given in Fig. 9.8.

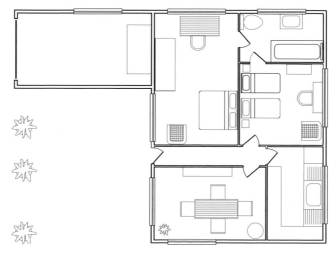

Fig. 9.8 Example – first method of inserting blocks

Example – second method of inserting blocks

1. Save the drawing which includes all the blocks to a suitable file name (e.g. building_symbols.dwg). Remember this drawing includes data of the blocks in its file.
2. *Left-click* **DesignCenter** in the **Tools/Palettes** panel (Fig. 9.9) or press the **Ctrl+2** keys. The **DesignCenter** palette appears on screen (Fig. 9.10).
3. With the outline plan (Fig. 9.5) on screen the symbols can all be *dragged* into position from the **DesignCenter**.

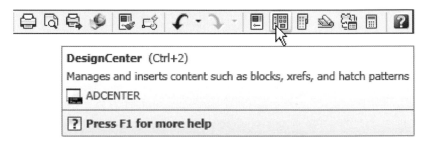

DesignCenter (Ctrl+2)

Manages and inserts content such as blocks, xrefs, and hatch patterns

ADCENTER

[?] **Press F1 for more help**

Fig. 9.9 Selecting **DesignCenter** from the **Standard Annotation** toolbar

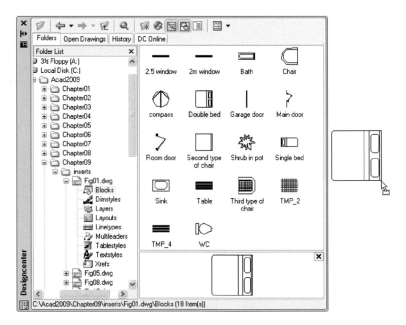

Fig. 9.10 The **DesignCenter** with the double bed block **dragged** on screen

Notes about **DesignCenter** palette

1. As with other palettes, the **DesignCenter** palette can be re-sized by *dragging* the palette to a new size from its edges or corners.
2. *Clicks* on one of the three icons at the top-right corner of the palette (Fig. 9.11) have the following results:
   - **Tree View Toggle** – changes from showing two areas – a **Folder List** and icons of the blocks within a file and icons of the blocks within a file – to a single area showing the block icons (Fig. 9.12).

- **Preview** – a *click* on the icon opens a small area at the base of the palette open showing an enlarged view of the selected block icon.
- **Description** – a *click* on the icon opens another small area with a description of the block.

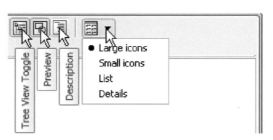

Fig. 9.11 The icons at the top-right corner of the **DesignCenter** palette

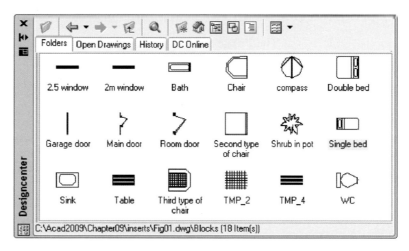

Fig. 9.12 The results of a click on **Tree View Toggle**

The **Explode** tool

Fig. 9.13 The **Explode** check box in the **Insert** dialog

A block is a single object no matter from how many objects it was originally constructed. This enables a block to be *dragged* about the drawing area as a single object.

A check box in the bottom-left-hand corner of the **Insert** dialog is labelled **Explode**. If a tick is in the check box, **Explode** will be set on and when a block is inserted it will be exploded into the objects from which it was constructed (Fig. 9.13).

Another way of exploding a block would be to use the **Explode** tool from the **Home/Modify** panel (Fig. 9.14). A *click* on the icon or *entering* **ex** at the command line brings prompts into the command line:

```
Command: _explode
Select objects: pick a block on screen 1 found
Select objects: right-click
Command:
```

and the *picked* object is exploded into its original objects.

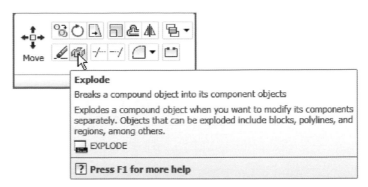

Fig. 9.14 The **Explode** tool icon in the **Home**/**Modify** panel

The **Purge** tool

The **Purge** tool can be called by *entering* **pu** at the command line or from the **Tools/Drawing Utilities** panel (Fig. 9.15).

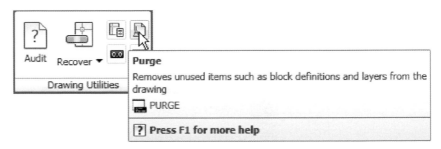

Fig. 9.15 Calling **Purge** from the **Tools**/**Drawing Utilities** panel

The **Purge** tool can be used to remove the data of blocks within a drawing, thus saving file space when a drawing which includes blocks is saved to disk.

To use the tool, in its dialog (Fig. 9.16) *click* the **Purge All** button and a sub-dialog appears naming a block to be purged. A *click* on the **Yes** button clears the data of the block from the drawing. Continue until all blocks that are to be purged are removed.

Take the drawing in Fig. 9.8 (p. 171) as an example. If all of the blocks are purged from the drawing, the file will be reduced from **145** kbytes to **67** kbytes when the drawing is saved to disk.

Using the **DesignCenter** (Fig. 9.19)

1. Construct the set of electric/electronic circuit symbols shown in Fig. 9.17 and make a series of blocks from each of the symbols.
2. Save the drawing to a file Fig17.dwg.

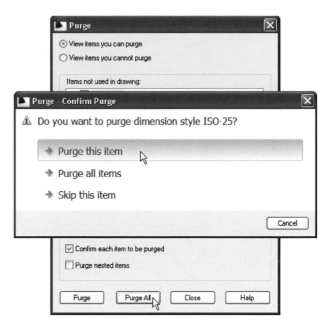

Fig. 9.16 The **Purge** dialog

3. Open the **acadiso.dwt** template. Open the **DesignCenter** with a *click* on its icon in the **Standard Annotation** toolbar.
4. From the **Folder list** select the file Fig17.dwg and *click* on **Blocks** under its file name. Then *drag* symbol icons from the **DesignCenter** into the drawing area as shown in Fig. 9.18. Ensure they are placed in appropriate positions in relation to each other to form a circuit. If necessary either **Move** or/and **Rotate** the symbols into correct positions.
5. Close the **DesignCenter** palette with a *click* on the **x** in the top-left-hand corner.
6. Complete the circuit drawing as shown in Fig. 9.19.

> **Note**
>
> Figure 9.19 does not represent an authentic electronics circuit.

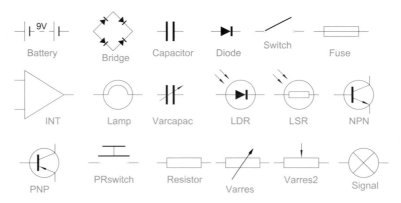

Fig. 9.17 Example using the **DesignCenter** – a set of electric/electronic systems

CHAPTER 9

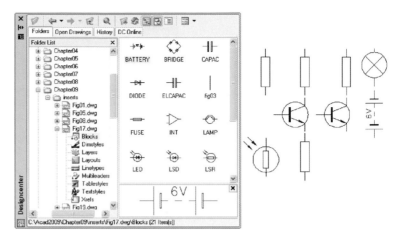

Fig. 9.18 Example using the **DesignCenter**

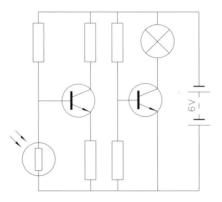

Fig. 9.19 Example using the **DesignCenter** – the completed circuit

Wblocks

Wblocks or written blocks are saved as drawing files in their own right and are not part of the drawing in which they have been saved.

Example – wblock (Fig. 9.20)

1. Construct a light emitting diode (**LED**) symbol and *enter* **w** at the command line. The **Write Block** dialog appears (Fig. 9.20).
2. *Click* the button marked with three full points (**…**) to the right of the **File name and path** field and from the **Browse for Drawing File** dialog which comes to screen select an appropriate directory. The directory name appears in the **File name and path** field. Add **LED. dwg** at the end of the name.
3. Make sure the **Insert units** is set to **Millimeters** in its pop-up list.
4. *Click* the **Select objects** button, Window the symbol drawing and when the dialog reappears, *click* the **Pick point** button, followed by selecting the left-hand end of the symbol.
5. Finally, *click* the **OK** button of the dialog and the symbol is saved in its selected directory as a drawing file **LED.dwg** in its own right.

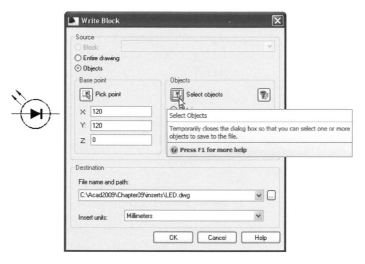

Fig. 9.20 Example – **Wblock**

Note on the DesignCenter

Drawings can be inserted into the AutoCAD window from the **DesignCenter** by dragging the icon representing the drawing into the window (Fig. 9.21).

When such a drawing is *dragged* into the AutoCAD window, the command line shows a sequence such as:

```
Command: _INSERT Enter block name or [?] <Fig26>:
  "Chapter07\inserts\Fig25.dwg"
Specify insertion point or [prompts]: pick
```

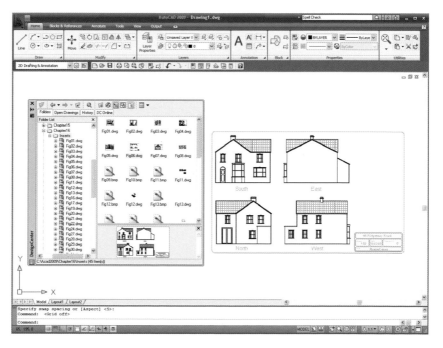

Fig. 9.21 An example of a drawing *dragged* from the **DesignCenter**

```
Enter X scale factor <1>: enter 0.4 right-click
Enter Y scale factor <use X scale factor>: right-click
Specify rotation angle <0>: right-click
Command:
```

REVISION NOTES

1. Blocks become part of the drawing file in which they were constructed.
2. Wblocks become drawing files in their own right.
3. Drawings or parts of drawings can be inserted in other drawings with the **Insert** tool.
4. Inserted blocks or drawings are single objects unless either the **Explode** check box of the **Insert** dialog is checked or the block or drawing is exploded with the **Explode** tool.
5. Drawings can be inserted into the AutoCAD drawing area using the **DesignCenter**.
6. Blocks within drawings can be inserted into drawings from the **DesignCenter**.

Exercises

Methods of constructing answers to the following exercises can be found in the free website:

http://books.elsevier.com/companions/9780750689830

1. Construct the building symbols shown in Fig. 9.22 in a drawing saved as **symbols.dwg**. Then using the **DesignCenter** construct a building drawing of the first floor of the house you are living in making use of the symbols. Do not bother too much about dimensions because this exercise is designed to practise using the idea of making blocks and using the **DesignCenter**.

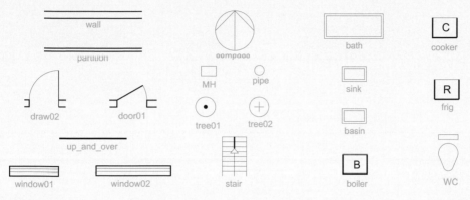

Fig. 9.22 Exercise 1

2. Construct drawings of the electric/electronics symbols in Fig. 9.17 (p. 175) and save them as blocks in a drawing file **electronics.dwg**.

3. Construct the electronics circuit given in Fig. 9.23 from the file **electronics.dwg** using the **DesignCenter**.

4. Construct the electronics circuit given in Fig. 9.24 from the file **electronics.dwg** using the **DesignCenter**.

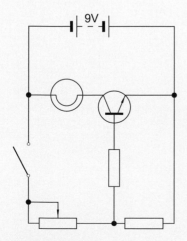

Fig. 9.23 Exercise 3

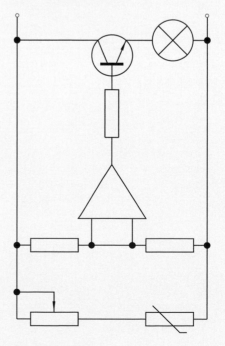

Fig. 9.24 Exercise 4

Other types of file format

AIMS OF THIS CHAPTER

The aims of this chapter are:
1. to introduce **Object Linking and Embedding (OLE)** and its uses;
2. to introduce the use of Encapsulated Postscript **(EPS)** files;
3. to introduce the use of Data Exchange Format **(DXF)** files;
4. to introduce raster files;
5. to introduce **Xrefs.**

Object Linking and Embedding

First example – Copying and Pasting (Fig. 10.2)

1. Open any drawing in the AutoCAD 2009 window (Fig. 10.1).
2. *Click* **Copy Link** in the **Home/Utilities** panel as indicated in Fig. 10.1.

Fig. 10.1 A drawing in the AutoCAD 2009 window showing **Copy Link** selected from the **Home/Utilities** panel

3. Open **Microsoft Word** and *click* on **Paste** in the **Edit** drop-down menu (Fig. 10.2). The drawing from the **Clipboard** appears in the **Microsoft Word** document (Fig. 10.2).
4. Add text as required.

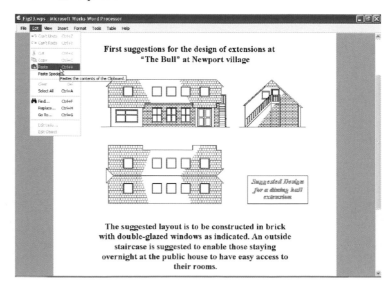

Fig. 10.2 Example – **Copying and Pasting**

Note

1. Similar results can be obtained using the **Copy Clip** and **Copy with Base Point** tools from the **Home/Utilities** panel of AutoCAD 2009.

2. The drawing could also be pasted back into the AutoCAD window – not that there would be much point in doing so, but anything in the **Clipboard** window can be pasted into any application.

Second example – EPS file (Fig. 10.5)

1. With the same drawing on screen *click* on **Export…** in the **Output/ Send** panel (Fig. 10.3). The **Export Data** dialog appears (Fig. 10.4). *Pick* **Encapsulated PS (*.eps)** from the **Files of type** pop-up list and then *enter* a suitable file name (**building.eps**) in the **File name field** and *click* the **Save** button.

2. Open a desktop publishing application. That shown in Fig. 10.5 is **PageMaker**.

Fig. 10.3 Selecting the **Export** tool icon from the **Output/Send** panel

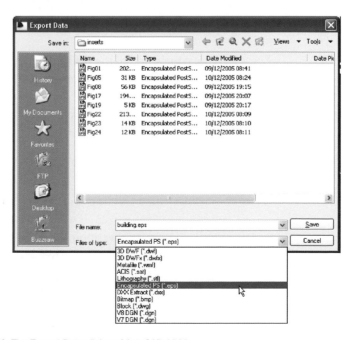

Fig. 10.4 The **Export Data** dialog of AutoCAD 2009

3. From the **File** drop-down menu *click* **Place...** A dialog appears listing files which can be placed in a PageMaker document. Among the files named will be **building.eps**. *Double-click* that file name and an icon appears the placing of which determines the position of the ***eps** file drawing in the PageMaker document (Fig. 10.5).
4. Add text as required.
5. Save the PageMaker document to a suitable file name.
6. Go back to the AutoCAD drawing and delete the title.
7. Make a new ***.eps** file with the same file name (**building.eps**).
8. Go back into **PageMaker** and *click* **Links Manager...** in the **File** drop-down menu. The **Links Manager** dialog appears (Fig. 10.6).

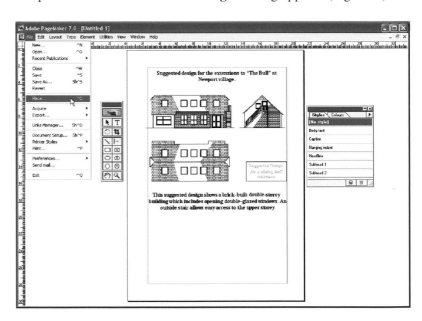

Fig. 10.5 An ***eps** file placed in position in a PageMaker document

Against the name of the **building.eps** file name is a dash and a note at the bottom of the dialog explaining that changes have taken place in the drawing from which the ***eps** had been derived. *Click the* **Update** button and when the document reappears the drawing in PageMaker no longer includes the erased title.

Note

1. This is **Object Linking and Embedding** (**OLE**). Changes in the AutoCAD drawing saved as an ***eps** file are linked to the drawing embedded in another application document, so changes made in the AutoCAD drawing are reflected in the PageMaker document.

2. There is actually no need to use the **Links Manager** because if the file from PageMaker is saved with the old *eps file in place, when it is reopened the file will have changed to the redrawn AutoCAD drawing, without the erased title.

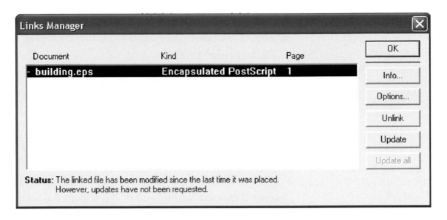

Fig. 10.6 The **Links Manager** dialog of PageMaker

DXF (Data Exchange Format) files

The ***.DXF** format was originated by Autodesk (publishers of AutoCAD), but is now in general use in most **CAD** (Computer Aided Design) software. A drawing saved to a *.dxf format file can be opened in most other CAD software applications. This file format is of great value when drawings are being exchanged between operators using different CAD applications.

Example – DXF file (Fig. 10.8)

1. Open a drawing in AutoCAD. This example is shown in Fig. 10.7.

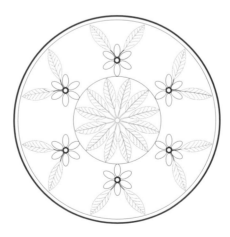

Fig. 10.7 Example – **DXF file**. Drawing to be saved as a dxf file

2. *Click* on **Save As...** in the **Menu Browser** dialog and in the **Save Drawing As** dialog which appears, *click* **AutoCAD 2007 DXF [*.dxf]** in the **Files of type** field pop-up list.

3. *Enter* a suitable file name. In this example this is **Fig07.** The extension **.dxf** is automatically included when the **Save** button of the dialog is *clicked* (Fig. 10.8).

4. The **DXF** file can now be opened in the majority of CAD applications and then saved to the drawing file format of the CAD in use.

> **Note**
>
> To open a **DXF** file in AutoCAD 2009, select **Open ...** from the **Menu Browser** dialog and in the **Select File** dialog select **AutoCAD 2007 DXF [*.dxf]** from the pop-up list from the **Files of type** field.

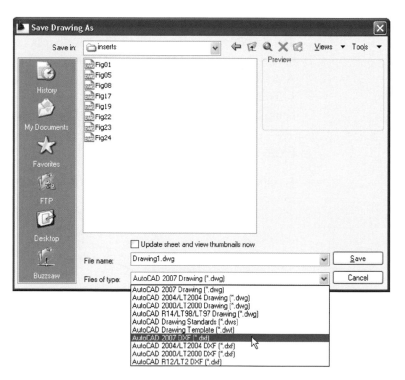

Fig. 10.8 The **Save Drawing As** dialog set up to save drawings in **DXF** format

Raster images

A variety of raster files can be placed into AutoCAD 2009 drawings from the **Select Image File** dialog brought to screen with a *click* on **Image** in the **Blocks & References/Reference** panel. In this example the selected raster file is a bitmap (extension ***.bmp**) of a rendered 3D model drawing constructed to the views in an assembly drawing of a lathe tool post (see Chapter 13 about the rendering of 3D models).

Example – placing a raster file in a drawing (Fig. 10.12)

1. *Click* **Image** in the **Blocks & References/Reference** panel (Fig. 10.9). The **Select Image File** dialog appears (Fig. 10.10). *Click* the file name of the image to be inserted, in this example **Fig. 44.bmp**. A preview appears in the **Preview** area of the dialog.
2. *Click* the **Open** button of the dialog. The **Image** dialog appears (Fig. 10.11).

Fig. 10.9 Selecting **Image** from the **Blocks and Reference/Reference** panel

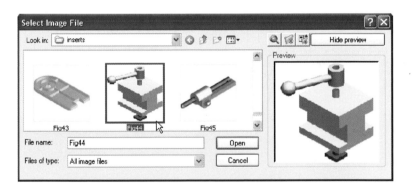

Fig. 10.10 The **Select Image File** dialog

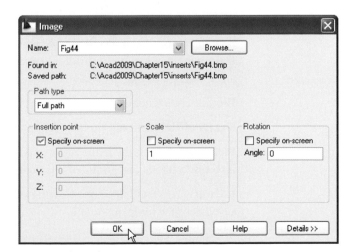

Fig. 10.11 The **Image** dialog

3. In the **Scale** field *enter* a suitable scale figure. The size the image will appear in the AutoCAD window can be seen with a *click* on the **Details** button which brings down an extension of the dialog that shows details about the resolution and size of the image.

4. *Click* the **OK** button, the command line shows:

```
Command: _imageattach

Specify insertion point <0,0>: the image
   attached to the cursor cross hairs appears pick
   a suitable point on screen

Command:
```

and the raster image appears at the *picked* point (Fig. 10.12).

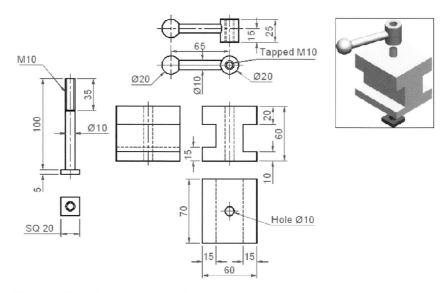

Fig. 10.12 Example – placing a raster file in a drawing

Note

As can be seen from the **Blocks & References/Block, Import** and **Data** panels, a variety of different types of. images can be inserted into an AutoCAD drawing. Some examples are:

Blocks – see Chapter 9.

External References (Xrefs) – see later in this chapter.

Field – A *click* on the name brings up the **Field** dialog. Practise inserting various categories of field names from the dialog.

Layout – A wizard appears allowing new layouts to be created and saved for new templates.

3D Studio – allows the insertion of images constructed in the Autodesk software **3D Studio** from files with the format ***.3ds**.

External References (Xrefs)

If a drawing is inserted into another drawing as an external reference, any changes made in the original Xref drawing are automatically reflected in the drawing into which the xref has been inserted.

Example – **External References** (Fig. 10.20)

1. Construct the three-view orthographic drawing given in Fig. 10.16. Dimensions of this drawing will be found on p. 301. Save the drawing to a suitable file name.
2. As a separate drawing construct Fig. 10.17. Save it as a wblock with the name of **Fig. 17.dwg** and with a base insertion point at one end of its centre line.
3. *Click* **External References** in the **View/Palettes** panel (Fig. 10.13). The **External Reference** palette appears (Fig. 10.14).

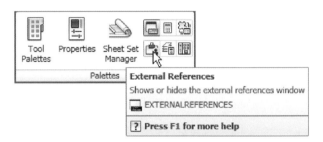

Fig. 10.13 Selecting **External Reference** from the **View/Palettes** panel

Fig. 10.14 The **External Reference** palette

4. *Click* its **Attach** button and select **Attach DWG…** from the pop-up list which appears when a *left-click* is held on the button. Select the drawing of a spindle (**Fig17.dwg**) from the **Select Reference file** dialog which appears followed by a *click* on the dialog's **Open** button. This brings up the **External Reference** dialog (Fig. 10.15) showing **Fig17** in its **Name** field. *Click* the dialog's **OK** button.

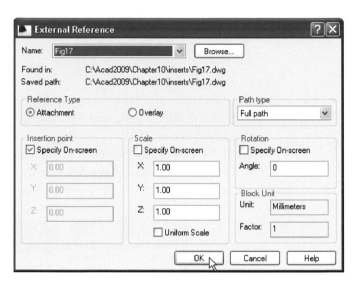

Fig. 10.15 The **External Reference** dialog

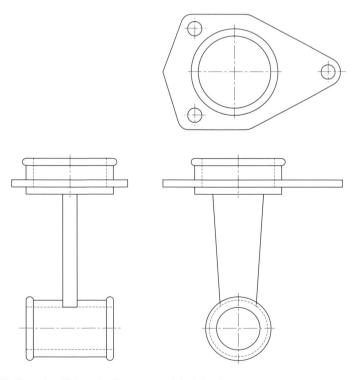

Fig. 10.16 Example – **External references** – original drawing

Fig. 10.17 The spindle drawing saved as **Fig17. dwg**

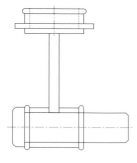

Fig. 10.18 The spindle in place in the original drawing

5. The spindle drawing appears on screen ready to be *dragged* into position. Place it in position as indicated in Fig. 10.18.
6. Save the drawing with its Xref to its original file name.
7. Open the drawing **Fig17.dwg** and make changes as shown in Fig. 10.19.
8. Now reopen the original drawing. The **External Reference** within the drawing has changed in accordance with the alterations to the spindle drawing.

> **Note**
>
> In this example to ensure accuracy of drawing the **External Reference** will need to be exploded and parts of the spindle changed to hidden detail lines.

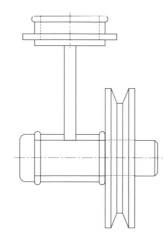

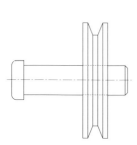

Fig. 10.19 The revised **spindle.dwg** drawing

Fig. 10.20 Example – **Xrefs**

Dgnimport and Dgnexport

Drawings constructed in MicroStation V8 format (***.dgn**) can be imported into AutoCAD 2009 format using the command **dgnimport** at the command line. AutoCAD drawings in AutoCAD 2004 format can be exported into MicroStation ***.dgn** format using the command **dgnexport**.

Example of importing a *.dgn drawing into AutoCAD

1. Figure 10.21 shows an example of an orthographic drawing constructed in MicroStation V8.
2. In AutoCAD 2009 at the command line *enter* **dgnimport**. The dialog (Fig. 10.22) appears on screen from which the required drawing file name can be selected. When the **Open** button of the dialog is *clicked* a

CHAPTER 10

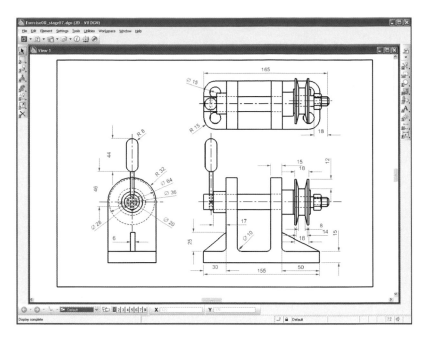

Fig. 10.21 Example – a drawing in MicroStation V8

warning window appears informing the operator of steps to take in order to load the drawing. When completed the drawing loads (Fig. 10.23).

In a similar manner AutoCAD drawing files can be exported to MicroStation using the command **dgnexport** *entered* at the command line.

Multiple Document Environment

1. Open several drawings in AutoCAD; in this example four separate drawings have been opened.

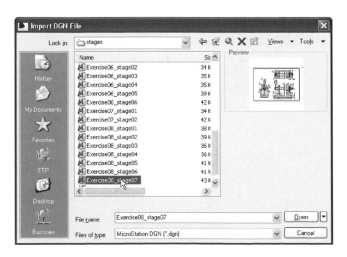

Fig. 10.22 The **DGN import File** dialog

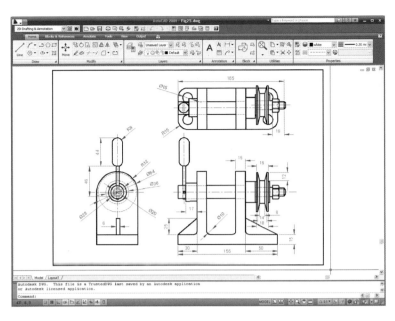

Fig. 10.23 The ***.dgn** file imported into AutoCAD 2009

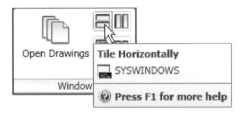

Fig. 10.24 Selecting **Tile Horizontally** from the **View/Window** panel

2. In the **View/Window** panel, *click* **Tile Horizontally** (Fig. 10.24). The four drawings rearrange as shown in Fig. 10.25.

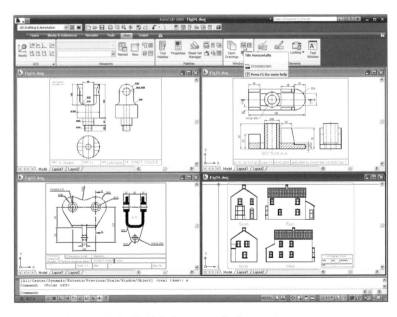

Fig. 10.25 Four drawings in the **Multiple Document Environment**

Note

The names of the drawings appear in the **Window** drop-down menu, showing their directories, file names and file name extensions.

REVISION NOTES

1. The **Edit** tools **Copy Clip, Copy with Base Point** and **Copy Link** enable objects from AutoCAD 2009 to be copied for **Pasting** onto other applications.
2. Objects can be copied from other applications to be pasted into the AutoCAD 2009 window.
3. Drawings saved in AutoCAD as **DXF** (***.dxf**) files can be opened in other computer-aided design (CAD) applications.
4. Similarly drawings saved in other CAD applications as ***.dxf** files can be opened in AutoCAD 2009.
5. **Raster** files of the format types ***.bmp, *.jpg, *pcx, *.tga, *.tif** among other raster type file objects can be inserted into AutoCAD 2009 drawings.
6. Drawings saved to the Encapsulated Postscript (***.eps**) file format can be inserted into documents of other applications.
7. Changes made in a drawing saved as an ***.eps** file will be reflected in the drawing inserted as an ***.eps** file in another application.
8. When a drawing is inserted into another drawing as an **External Reference** changes made to the inserted drawing will be updated in the drawing into which it has been inserted.
9. A number of drawings can be opened in the AutoCAD 2009 window.
10. Drawings constructed in MicroStation V8 can be imported into AutoCAD 2009 using the command **dgnimport.**
11. Drawings constructed in AutoCAD 2009 can be saved as MicroStation ***.dgn** drawings to be opened in MicroStation V8.

Exercises

Methods of constructing answers to the following exercises can be found in the free website:

http://books.elsevier.com/companions/9780750689830

1. Figure 10.26 shows a pattern formed by inserting an **External Reference** and then copying or arraying the **External Reference**. The hatched parts of the **External Reference** drawing were then changed using a different hatch pattern. The result of the change is shown in Fig. 10.27. Construct a similar **Xref** drawing, insert as an **Xref**, array or copy to form the pattern, then change the hatching, save the **Xref** drawing and note the results.

the roller and save to a file name **roller.dwg**. Then as a separate drawing construct a front view of the two end holders in their correct positions to receive the roller and save to the file name **assembly.dwg**. Insert the roller drawing into the assembly drawing as an **Xref**. Open the **roller.dwg** and change its outline as shown in Fig. 10.30. Save the drawing. Open the **assembly.dwg** and note the change in the inserted **Xref**.

Fig. 10.26 Exercise 1 – original pattern

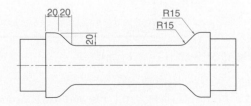

Fig. 10.28 Exercise 2 – a rendering of the holders and roller

Fig. 10.27 Exercise 1

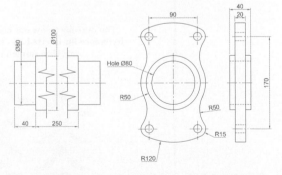

Fig. 10.29 Exercise 2 – details of the parts of the roller and holders

2. Figure 10.28 is a rendering of a roller between two end holders. Figure 10.29 gives details of the end holders and the roller in orthographic projections. Construct a full-size front view of

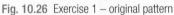

Fig. 10.30 The amended **Xref** drawing

CHAPTER 10

3.   *Click* **Image…** in the **Reference** panel and insert a **JPEG** image (***.jpg** file) of a photograph into the
AutoCAD 2009 window. An example is given in Fig. 10.31.

Fig. 10.31 Exercise 3 – example

4.   Using **Copy** from the **Insert** drop-down menu, copy a drawing from AutoCAD 2009 into a Microsoft Word
document. An example is given in Fig. 10.32. Add some appropriate text.

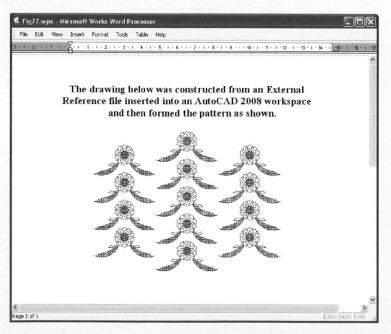

Fig. 10.32 Exercise 4 – an example

5.   The plan in Figs 10.1, 10.2 and 10.3 is incorrect in that some details have been missed from the drawing.
Can you identify the error?

Chapter 11

Sheet sets

AIMS OF THIS CHAPTER

The aims of this chapter are:

1. to introduce sheet sets;
2. to describe the use of the **Sheet Set Manager**;
3. to give an example of a sheet set based on a the design of a two-storey house.

Sheet sets

When anything is to be manufactured or constructed, whether it be a building, an engineering design, an electronics device or any other form of manufactured artefact, a variety of documents, many in the form of technical drawings, will be needed to convey to those responsible for constructing the design all the information necessary to be able to proceed according to the wishes of the designer. Such sets of drawings may be passed between the people or companies responsible for the construction, enabling all those involved to make adjustments or suggest changes to the design. In some cases there may well be a considerable number of drawings required in such sets of drawings. In AutoCAD 2009 all of the drawings from which a design is to be manufactured can be gathered together in a **sheet set**. This chapter shows how a much-reduced sheet set of drawings for the construction of a house at 62 Pheasant Drive can be produced. Some other drawings, particularly detail drawings, would be required in this example, but to save page space, the sheet set described here consists of only four drawings with a subset of another four drawings.

A sheet set for 62 Pheasant Drive

1. Construct a template **62 Pheasant Drive.dwt** based upon the **acadiso.dwt** template, but including a border and a title block. Save the template in a **Layout1** format. An example of the title block from one of the drawings constructed in this template is shown in Fig. 11.1.

Fig. 11.1 The title block from Drawing number **2** of the sheet set

2. Construct each of the drawings which will form the sheet set in this drawing template. The whole set of drawings is shown in Fig. 11.2. Save the drawings in a directory – in this example this has been given the name **62 Pheasant Drive**.
3. *Click* **Sheet Set Manager** in the **View/Palettes** panel (Fig. 11.3). The **Sheet Set Manager** palette appears (Fig. 11.4). *Click* **New Sheet Set...** in the pop-up menu at the top of the palettes. The first of a series of **Create Sheet Set** dialogs appears – the **Create Sheet Set – Begin** dialog (Fig. 11.5). *Click* the radio button next to **Existing drawings**, followed by a *click* on the **Next** button and the next dialog **Sheet Set Details** appears (Fig. 11.6).

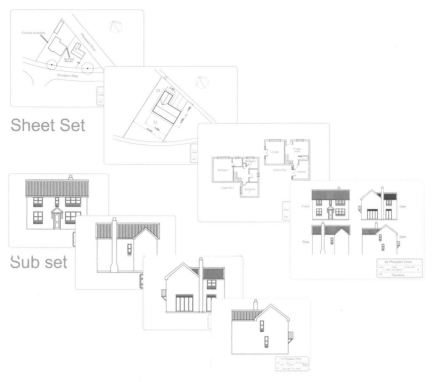

Fig. 11.2 The eight drawings in the **62 Pheasant Drive** sheet set

Fig. 11.3 Selecting **Sheet Set Manager** from the **View/Palettes** panel

Fig. 11.4 The **Sheet Set Manager** palette

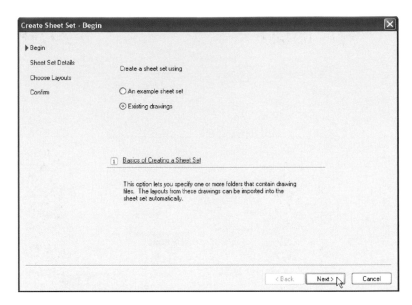

Fig. 11.5 The first of the **Create Sheet Set** dialogs – **Begin**

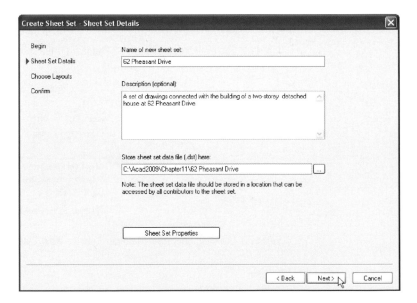

Fig. 11.6 The **Sheet Set Details** dialog

4. *Enter* details as shown in the dialog as shown in Fig. 11.6. Then *click* the **Next** button to bring the **Choose Layouts** dialog to screen (Fig. 11.7).
5. *Click* its **Browse** button and from the **Browse for Folder** list which comes to screen, *pick* the directory **62 Pheasant Drive**. *Click* the **OK** button and the drawings held in the directory appears in the **Choose Layouts** dialog (Fig. 11.7. If satisfied the list is correct, *click* the **Next** button. A **Confirm** dialog appears (Fig. 11.8). If satisfied *click* the **Finish** button and the **Sheet Set Manager** palette appears, showing the drawings which will be in the **62 Pheasant Drive** sheet set (Fig. 11.9).

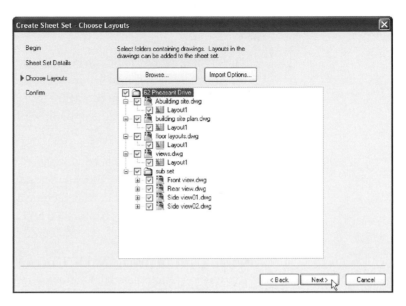

Fig. 11.7 The **Choose Layouts** Dialog

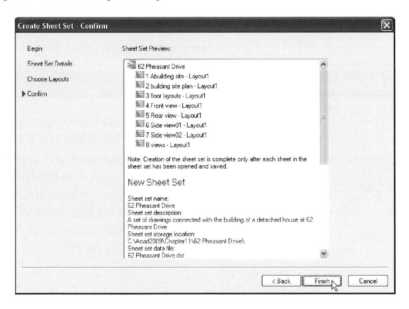

Fig. 11.8 The **Confirm** dialog

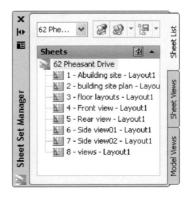

Fig. 11.9 The **Sheet** Manager palette for **62 Pheasant Drive**

Note

1. The eight drawings in the sheet set are shown in Fig. 11.9. If any of the drawings in the sheet set are subsequently amended or changed, when the drawings is opened again from the **62 Pheasant Drive Sheet Manager** palette, the drawing will include any changes or amendments.

2. Drawings can only be placed into sheet sets if they have been saved in a **Layout** screen. Note that all the drawings shown in the **62 Pheasant Drive** Sheet Set Manager have **Layout1** after the drawing names because each has been saved after being constructed in a **Layout1** template.

3. Sheet sets in the form of **DWF** (Design Web Format) files can be sent via email to others who are using the drawings or placed on an intranet. The method of producing a **DWF** for the **62 Pheasant Drive** sheet set follows.

62 Pheasant Drive DWF

1. In the **62 Pheasant Drive** Sheet Set Manager *click* the **Publish to DWF** icon (Fig. 11.10). The **Select DWF File** dialog appears (Fig. 11.11). *Enter* **62 Pheasant Drive** in the **File name** field followed by a *click* on the **Select** button. The **Publish Job in Progress** icon in the bottom right-hand corner of the AutoCAD 2009 window starts fluctuating in shape, showing that the DWF file is being processed (Fig. 11.12). When the icon becomes stationary, *right-click* the icon and *click* **View DWF file** in the *right-click* menu which appears (Fig. 11.13).

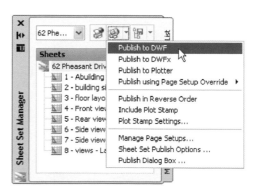

Fig. 11.10 The **Publish to DWF** icon in the **Sheet Set Manager**

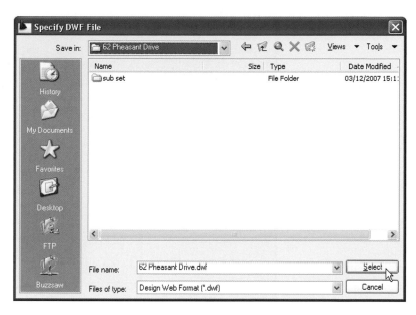

Fig. 11.11 The **Select DWF File** dialog

Fig. 11.12 The **Publish Job in Progress** icon

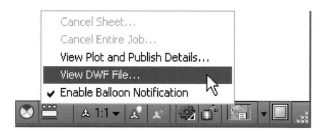

Fig. 11.13 The *right-click* menu of the icon

2. The **Autodesk DWF Viewer** window appears showing the **62 Pheasant Drive.dwf** file (Fig. 11.14). *Click* in any of the icons of the thumbnails of the drawings in the viewer and the drawing appears in the right-hand area of the viewer.

3. If required the DWF Viewer file can be sent between people by email as an attachment, opened in a company's intranet or, indeed, included within an internet web page.

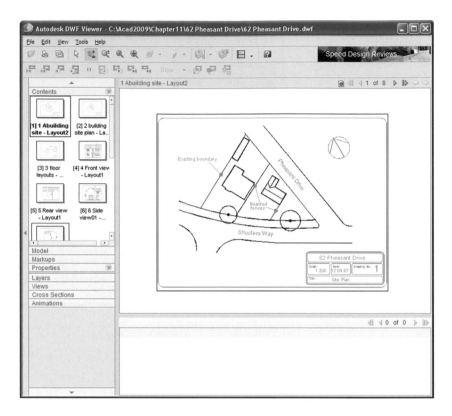

Fig. 11.14 The **Autodesk DWF Viewer** showing details of the **62 Pheasant Drive.dwf** file

REVISION NOTES

1. To start off a new sheet set, select the **Sheet Set Manager** icon in the **Tools/Palettes** panel.
2. Sheet sets can only contain drawings saved in **Layout** format.
3. Sheet sets can be published as **Design Web Format** (***.dwf**) files which can be sent between offices by email, published on an intranet or published on a web page.
4. Sub-sets can be included in sheet sets.
5. Changes or amendments made to any drawings in a sheet set are reflected in the sheet set drawings when the sheet set is opened.

Exercises

Methods of constructing answers to the following exercises can be found in the free website:

http://www.books.elsevier.com/companions/9780750689830

1. Figure 11.15 shows an exploded orthographic projection of the parts of a piston and its connecting rod. There are four parts in the assembly. Small drawings of the required sheet set are shown in Fig. 11.17. Construct the drawing shown in Fig. 11.15 and also the four drawings of its parts. Save each of the drawings in a **Layout1** format and construct the sheet set which contains the five drawings.

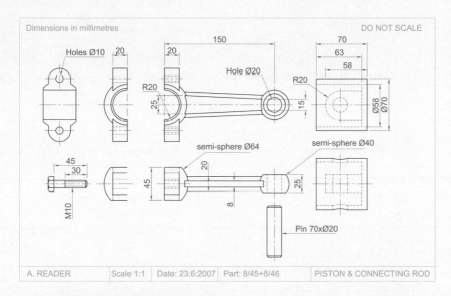

Fig. 11.15 Exercise 1 – exploded orthographic projection

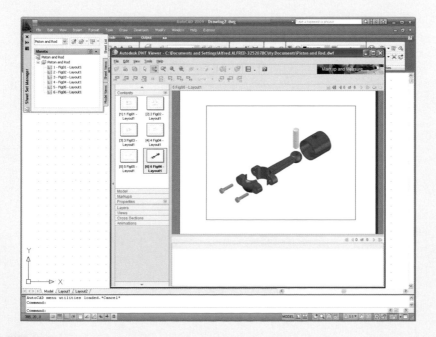

Fig. 11.16 The **DWF** for Exercise 1

Construct the **DWF** file of the sheet set. Experiment sending it to a friend via email as an attachment to a document, asking him/her to return the whole email to you without changes. When the email is returned, open its DWF file and *click* each drawing icon in turn to check the contents of the drawings.

2. Construct a similar sheet set as in the answer to Exercise 1 from the exploded orthographic drawing of a **Machine adjusting spindle** given in Fig. 11.18.

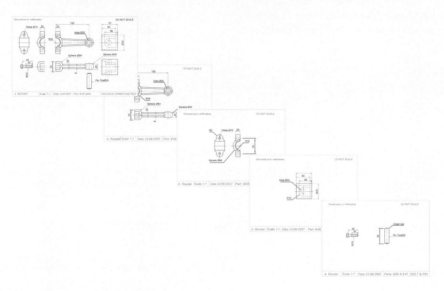

Fig. 11.17 Exercise 1 – The five drawings in the sheet set

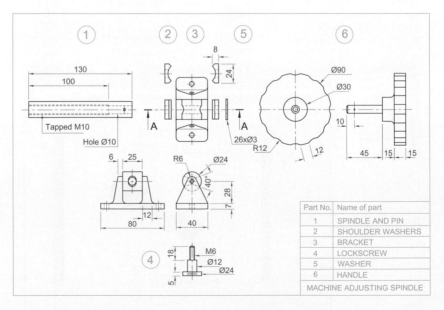

Fig. 11.18 Exercise 2

Part No.	Name of part
1	SPINDLE AND PIN
2	SHOULDER WASHERS
3	BRACKET
4	LOCKSCREW
5	WASHER
6	HANDLE
MACHINE ADJUSTING SPINDLE	

Part 2

3D Design

Introducing 3D modelling

AIMS OF THIS CHAPTER

The aims of this chapter are:

1. to introduce the tools used for the construction of 3D solid models;
2. to give examples of the construction of 3D solid models using tools from the **Home/3D Modeling** panel;
3. to give examples of 2D outlines suitable as a basis for the construction of 3D solid models;
4. to give examples of constructions involving the Boolean operators – **Union, Subtract** and **Intersect**.

Introduction

As shown in Chapter 1, the AutoCAD coordinate system includes a third coordinate direction **Z**, which, when dealing with 2D drawing in previous chapters, has not been used. 3D model drawings make use of this third **Z** coordinate.

The **3D Modeling** workspace

It is possible to construct 3D model drawings in the **AutoCAD Classic** or **2D Drafting & Annotation** workspaces, but in Part 2 of this book we will be working in the **3D Modeling** workspace. To set this workspace *click* the **Workspace Settings** icon in the status bar and select **3D Modeling** from the menu which appears (Fig. 12.1). The **3D Modeling** workspace appears (Fig. 12.2).

Fig. 12.1 Selecting **3D Modeling** from the **Workspace Switching** menu in the status bar

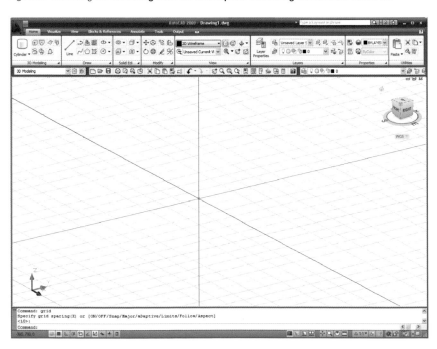

Fig. 12.2 The **3D Modeling** workspace in **Parallel projection**

The workspace in Fig. 12.2 shows grid lines in **Parallel projection** mode, brought about by *entering* **perspective** at the command line, followed by *entering* **0** in response to the prompt which appears. This is the window in which the examples in this chapter will be constructed. Note the **ViewCube** at the top-right-hand corner of the drawing area. In Fig. 12.2 this has been highlighted by moving the cursor onto the cube. Changes can be made to the appearance and uses of the cube in the **ViewCube Settings** dialog brought to screen from the *right-click* menu of the **ViewCube** (Fig. 12.3).

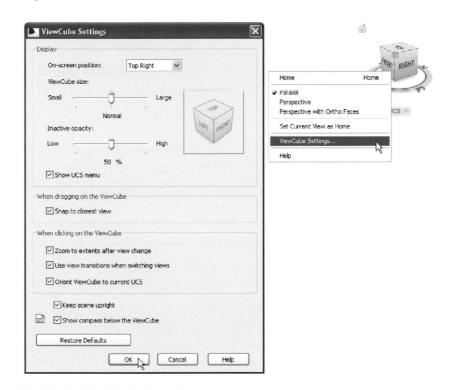

Fig. 12.3 The **ViewCube Settings** dialog

Methods of calling tools for 3D modeling

When calling the tools for the construction of 3D model drawings, similar methods can be used as when constructing 2D drawings:

1. A *click* on a tool icon in the **Home/3D Modeling** panel (Fig. 12.4).
2. A *click* on a tool icon in the **Modeling** toolbar.
3. A *click* on the name of a tool from the **Draw/Modeling** drop-down menu brings the tool into action.
4. *Entering* the tool name at the command line in the command window, followed by pressing the *Return* button of the mouse or the **Return** key of the keyboard, brings the tool into action.
5. Some of the 3D tools have an abbreviation which can be *entered* at the command line instead of its full name.

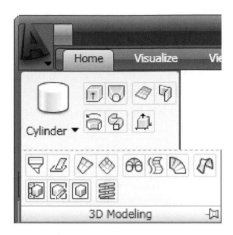

Fig. 12.4 The 3D tool icons in the **Home**/**3D Modeling** panel

Note

1. As when constructing 2D drawings, no matter which method is used – and most operators will use a variety of these five methods – the result of calling a tool consists in prompt sequences appearing at the command prompt as in the following example:

```
Command: enter box right-click
Specify first corner or [Center]: enter 90,120
right-click
Specify other corner or [Cube/Length]: enter
150,200
Specify height or [2Point]: enter 50
or, if the tool is called from its tool icon, or
from a drop-down menu:
Command: _box
Specify first corner or [CEnter]: enter 90,120
right-click
Specify other corner or [Cube/Length]: enter
150,200
Specify height or [2Point]: enter 50
```

2. In the following pages, if the tool's sequences are to be repeated, they may be replaced by an abbreviated form such as:

```
Command: box
[prompts]: 90,120
[prompts]: 150,200
```

The **Polysolid tool** – Fig. 12.8

Fig. 12.5 *Click* **Top** in the **ViewCube**

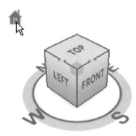

Fig. 12.6 Selecting **Isometric** from the **ViewCube**

1. Make sure layer **0** is current.
2. *Click* **Top** in the **ViewCube** (Fig. 12.5). The screen switches to a **Top** view.
3. Construct an octagon of edge length **60** using the **Polygon** tool.
4. Set layer **0** current and *click* the house icon in the **ViewCube** (Fig. 12.6). The screen switches to an **Isometric** view.
5. Call the **Polysolid** tool with a *click* on its tool icon in the **3D Modeling** panel (Fig. 12.7). The command line shows:

```
Command: _Polysolid Height = 0, Width = 0,
Justification = Center
Specify start point or [Object/Height/Width/
Justify] <Object>: enter h (Height)
Specify height <4>: 60
Height = 60, Width = 0, Justification = Center
Specify start point or [Object/Height/Width/
Justify] <Object>: enter w (Width)
Specify width <0>: 5
Height = 60, Width = 5, Justification = Center
Specify start point or [Object/Height/Width/
Justify] <Object>:
Specify next point or [Arc/Undo]: pick one corner
of octagon
Specify next point or [Arc/Undo]: pick second
corner
```

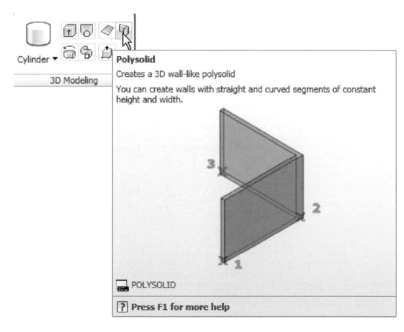

Fig. 12.7 The **Polysolid** tool icon in the **Home/3D Modeling** panel

```
Specify next point or [Arc/Close/Undo]: pick third
Specify next point or [Arc/Close/Undo]: pick
  fourth
Specify next point or [Arc/Close/Undo]: pick fifth
Specify next point or [Arc/Close/Undo]: pick sixth
Specify next point or [Arc/Close/Undo]: pick
  seventh
Specify next point or [Arc/Close/Undo]: pick last
Specify next point or [Arc/Close/Undo]: enter c
  (Close)
Command:
```

Fig. 12.8 Example Polysolid

and the **Polyline** forms (Fig. 12.8).

2D outlines suitable for 3D models

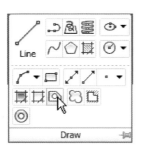

Fig. 12.9 Selecting the **Region** tool from the **Home/Draw** panel

When constructing 2D outlines suitable as a basis for constructing some forms of 3D model, select a tool from the **Home/Draw** panel, or *enter* tool names or abbreviations for the tools at the command line. If constructed using tools such as **Line**, **Circle** and **Ellipse**, before being of any use for 3D modelling, outlines must be changed into regions with the **Region** tool. Closed polylines can be used without the need to use the **Region** tool.

First example – **Line** outline and **Region** (Fig. 12.10)

1. Construct the left-hand drawing of Fig. 12.7 using the **Line** tool.
2. *Click* the **Region** tool from the **Home/Draw** panel (Fig. 12.9), or *enter* **reg** at the command line.

```
Command: _region
Select objects: window the drawing 12 found
Select objects: right-click
1 Region created
Command:
```

and the **Line** outline is changed to a region – right-hand drawing of Fig. 12.10.

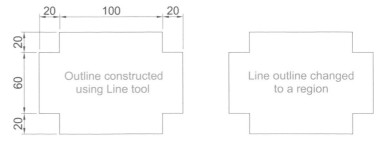

Fig. 12.10 First example – **Line** outline and **Region**

Second example – **Union**, **Subtract** regions (Fig. 12.11)

1. In the **ViewCube/Top** view, construct drawing **1** of Fig. 12.11 and with the **Copy** tool (**Home/Modify** panel), copy the drawing three times to produce drawings **2**, **3** and **4**.
2. With the **Region** tool change all the outlines into regions.
3. Drawing **2** – call the **Union** tool from the **Home/Solid Editing** panel (Fig. 12.12). The command line shows:

```
Command: _union
Select objects: pick the left-hand region 1 found
Select objects: pick the circular region 1 found,
  2 total
Select objects: pick the right-hand region 1
  found, 3 total
Command:
```

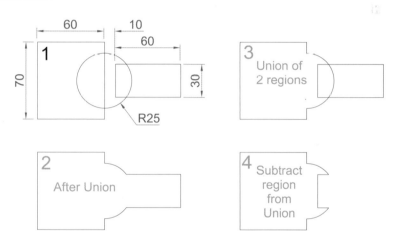

Fig. 12.11   Second example – **Union, Subtract** and **Region**

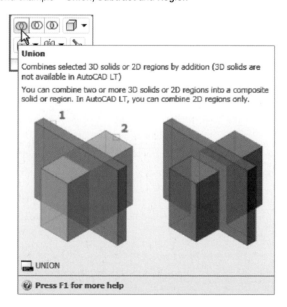

Fig. 12.12   The **Union** tool icon in the **Home/Solid Editing** panel

4. Drawing **3** – with the **Solid, Union** tool form a union of the left-hand region and the circular region.

5. Drawing **4** – call the **Subtract** tool, also from the **Home/Solid Editing** panel. The command line shows:

```
Command: _subtract Select solids and regions to
    subtract from...
Select objects: pick the region just formed 1
    found
Select objects: right-click
Select solids and regions to subtract: pick the
    right-hand region 1 found
Select objects: right-click
Command:
```

Third example – **Intersect** and **Region** (Fig. 12.13)

1. Construct drawing **1** of Fig. 12.13.
2. With the **Region** tool, change the three outlines into regions.
3. With the **Copy** tool, copy the three regions.
4. Drawing **2** – call the **Intersect** tool from the **Home/3D Modeling** panel. The command line shows:

```
Command: _intersect
Select objects: pick one of the circles 1 found
Select objects: pick the other circle 1 found, 2
total
Select objects: right-click
Command:
```

and the two circular regions intersect with each other to form a region.

5. Drawing **3** – repeat using the **Intersect** tool from the **Home/Solid Editing** panel on the intersection of the two circles and the rectangular region.

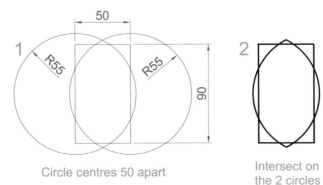

Fig. 12.13 Third example – **Intersect** and **Region**

The Extrude tool

The **Extrude** tool can be called with a *click* on its tool icon in the
Home/3D Modeling panel (Fig. 12.14), from the **Modeling** toolbar, from
the **Draw** drop-down menu, or by *entering* **extrude** or its abbreviation **ext**
at the command line.

> **Note**
>
> In this chapter, 3D models are shown in illustrations as they appear in
> the **acadiso3D.dwt** template screen. In later chapters, 3D models are
> sometimes shown in outline only. This is to allow the reader to see the
> parts of 3D models in future chapters more clearly in the illustrations.

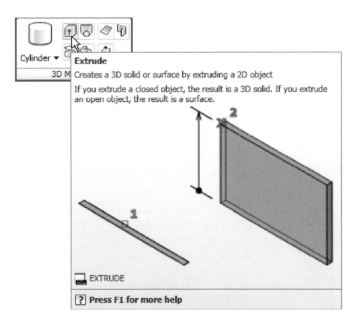

Fig. 12.14 The **Extrude** tool from the **Home/3D Modeling** panel

Examples of the use of the **Extrude** tool

The three examples of forming regions given in Figs 12.10, 12.11 and
12.13 are used here to show results of using the **Extrude** tool.

First example – **Extrude** (Fig. 12.15)

From the first example of forming a region:

1. Call the **Extrude** tool (Fig. 12.14). The command line shows:

```
Command: _extrude
Current wire frame density: ISOLINES = 4
```

```
Select objects to extrude: pick region 1 found
Select objects to extrude: right click
Specify height of extrusion or [Direction/Path/
  Taper angle] <45>: enter 50 right-click
Command:
```

2. Select **ViewCube/Isometric**. The extrusion appears in an isometric view.
3. Call **Zoom** and zoom to **1**.

Fig. 12.15 First example – **Extrude**

Fig. 12.16 Second example – new position of the **ViewCube**

Fig. 12.17 Second example – **Extrude**

> **Note**
>
> 1. In the above example we made use of an isometric view possible from the **ViewCube** (Fig. 12.5). The **ViewCube** can be manipulated to show a variety of views by *dragging* it to its required positions under mouse control.
> 2. Note the **Current wire frame density**: ISOLINES = 4 in the prompts sequence when Extrude is called. The setting of 4 is suitable when extruding plines or regions consisting of straight lines, but when arcs are being extruded it may be better to set ISOLINES to a higher figure as follows:
>
> ```
> Command: enter isolines right-click
> Enter new value for ISOLINES <4>: enter 16
> right-click
> Command:
> ```

Second example – Extrude (Fig. 12.17)

From the second example of forming a region:

1. Set **ISOLINES** to **16**.
2. Call the Extrude tool. The command line shows:

```
Command: _extrude
Current wire frame density: ISOLINES = 16
Select objects to extrude: pick the region 1 found
Select objects to extrude: right-click
Specify height of extrusion or [Direction/Path/
  Taper angle]: enter t right-click
Specify angle of taper for extrusion: enter 5
  right-click
Specify height of extrusion or [Direction/Path/
  Taper angle]: enter 100 right-click
Command:
```

3. *Drag* the **ViewCube** to a new view as shown in Fig. 12.16.
4. **Zoom** to **1**.

The result is shown in Fig. 12.17.

Third example – **Extrude** (Fig. 12.18)

From the third example of forming a region:

1. Place the screen in the **ViewCube/Front** view.
2. Construct a semicircular arc from the centre of the region.
3. Place the screen in the **ViewCube/Top**.
4. With the **Move** tool, move the arc to the centre of the region.
5. Place the screen in the **ViewCube/Isometric** view.
6. Set **ISOLINES** to **24**.
7. Call the **Extrude** tool. The command line shows:

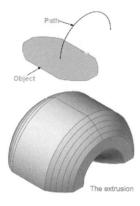

```
Command: _extrude
Current wire frame density: ISOLINES = 24
Select objects to extrude: pick the region 1
   found
Select objects to extrude: right-click
Specify height of extrusion or [Direction/Path/
   Taper angle] <100>: enter p right-click
Select extrusion path or [Taper angle]: pick the
   path
Command:
```

The result is shown in Fig. 12.18.

Fig. 12.18 Third example –
Extrude

The **Revolve** tool

The **Revolve** tool can be called with a *click* on its tool icon in the **Modeling** toolbar, by a *click* on its tool icon in the **Home/3D Modeling** panel, by a *click* on its name in the **Modeling** sub-menu of the **Draw** drop-down menu, or by *entering* **revolve** at the command line, or its abbreviation **rev**.

Examples of the use of the **Revolve** tool

Solids of revolution can be constructed from closed plines or from regions.

First example – **Revolve** (Fig. 12.21)

1. Construct the closed polyline shown in Fig. 12.19.
2. Set **ISOLINES** to **24**.
3. Call the **Revolve** tool from the **Home/3D Modeling** control panel (Fig. 12.20). The command line shows:

```
Command: _revolve
Current wire frame density: ISOLINES = 24
```

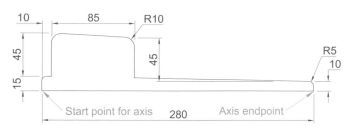

Fig. 12.19 First example – **Revolve**. The closed pline

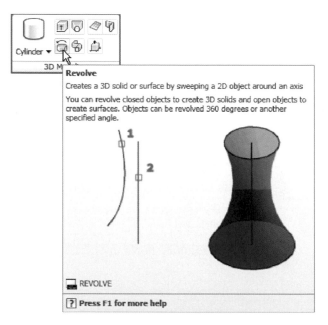

Fig. 12.20 The **Revolve** tool from the **Home**/**3D Modeling** panel

Fig. 12.21 First example – **Revolve**

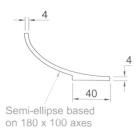

Fig. 12.22 Second example – **Revolve**. The pline outline

```
Select objects to revolve: pick the polyline 1
found
Select objects to revolve: right-click
Specify axis start point or define axis by [Object/
X/Y/Z] <Object > : pick
Specify axis endpoint: pick
Specify angle of revolution or [Start angle]
<360>: right-click
Command:
```

4. Place in the **ViewCube**/**Isometric** view. **Zoom** to **1**.

The result is shown in Fig. 12.21.

Second example – **Revolve** (Fig. 12.23)

1. Place the screen in the **ViewCube**/**Front** view. **Zoom** to **1**.
2. Construct the pline outline (Fig. 12.22).
3. Set **ISOLINES** to **24**.

4. Call the **Revolve** tool and construct a solid of revolution.
5. Place the screen in the **ViewCube/Isometric** view. **Zoom** to **1**.

Third example – **Revolve** (Fig. 12.24)

1. Construct the pline (left-hand drawing of Fig. 12.24). The drawing must be either a closed pline or a region.
2. Call **Revolve** and form a solid of revolution through 180°.
3. Place the model in the **ViewCube/Isometric** view. **Zoom** to **1**.

The result is shown in the right-hand side of Fig. 12.24.

Fig. 12.23 Second example – **Revolve**

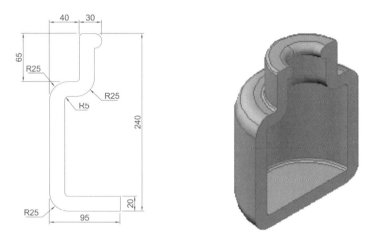

Fig. 12.24 Third example – **Revolve**. The outline to be revolved and the solid of the revolution

Other tool from the **Home/3D Modeling** panel

First example – **Box** (Fig. 12.26)

1. Place the window in the **ViewCube/|Front** view.
2. *Click* the **Box** tool icon in the **Home/3D Modeling** panel (Fig. 12.25). The command line shows:

```
Command: _box
Specify first corner or [Center]: enter 90,90
  right-click
Specify other corner or [Cube/Length]: enter
  110,-30 right-click
Specify height or [2Point]: enter 75 right-click
Command: right-click
BOX Specify first corner or [Center]: 110,90
Specify other corner or [Cube/Length]: 170,70
Specify height or [2Point]: 75
```

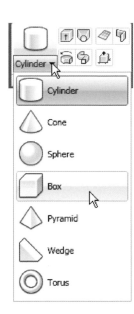

Fig. 12.25 Selecting the **Box** tool from the pop-up in the **Home**/**3D Modeling** panel

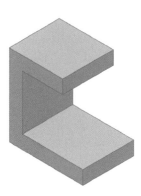

Fig. 12.26 First example – **3D Objects**

```
Command:
BOX Specify first corner or [Center]: 110,-10
Specify other corner or [Cube/Length]: 200,-30
Specify height or [2Point]: 75
Command:
```

3. Place in the **ViewCube/Isometric** view. **Zoom** to **1**.
4. Call the **Union** tool from the **Home/Solid Editing** panel. The command line shows:

```
Command: _union
Select objects: pick one of the boxes 1 found
Select objects: pick the second of box 1 found, 2
   total
Select objects: pick the third box 1 found, 3
   total
Select objects: right-click
Command:
```

and the three boxes are joined in a single union. See Fig. 12.26.

Second example – **Sphere** and **Cylinder** (Fig. 12.27)

1. Set **ISOLINES** to **16**.
2. *Click* the **Sphere** tool icon from the **Home/3D Modeling** panel. The command line shows:

```
Command: _sphere
Specify center point or [3P/2P/Ttr]: 180,170
Specify radius or [Diameter]: 50
Command:
```

3. *Click* the **Cylinder** tool icon in the **Home/3D Modeling** panel. The command line shows:

```
Command: _cylinder
Specify center point of base or [3P/2P/Ttr/
   Elliptical]: 180,170
Specify base radius or [Diameter]: 25
Specify height or [2Point/Axis endpoint]: 110
Command:
```

4. Place the screen in the **ViewCube/Front** view. **Zoom** to **1**.
5. With the **Move** tool (from the **Home/Modify** panel), move the cylinder vertically down so that the bottom of the cylinder is at the bottom of the sphere.
6. *Click* the **Subtract** tool icon in the **Home/Solid Editing** panel. The command line shows:

```
Command: _subtract Select solids and regions to
   subtract from...
Select objects: pick the sphere 1 found
```

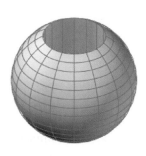

```
Select objects: right-click
Select solids and regions to subtract
Select objects: pick the cylinder 1 found
Select objects: right-click
Command:
```

7. Place the screen in the **ViewCube/Isometric** view. **Zoom** to **1**.

The result is shown in Fig. 12.27.

Fig. 12.27 Second
example – **3D Objects**

Third example – **Cylinder**, **Cone** and **Sphere** (Fig. 12.28)

1. Call the **Cylinder** tool and with a centre 170,150 construct a cylinder of radius **60** and height **15**.
2. *Click* the **Cone** tool in the **Home/3D Modeling** panel. The command line shows:

```
Command: _cone
Specify center point of base or [3P/2P/Ttr/
  Elliptical]: 170,150
Specify base radius or [Diameter]: 40
Specify height or [2Point/Axis endpoint/Top
  radius]: 150
Command:
```

3. Call the **Sphere** tool and construct a sphere of centre **170,150** and radius **45**.
4. Place the screen in the **Front** view and with the **Move** tool, move the cone and sphere so that the cone is resting on the cylinder and the centre of the sphere is at the apex of the cone.
5. Place in the **ViewCube/Isometric** view, **Zoom** to **1** and with **Union** form a single 3D model from the three objects.

The result is shown in Fig. 12.28.

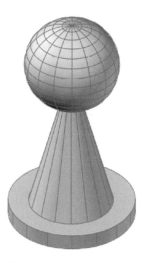

Fig. 12.28 Third example –
3D Objects

Fourth example – **Box** and **Wedge** (Fig. 12.29)

1. *Click* the **Box** tool icon in the **Home/3D Modeling** panel and construct two boxes, the first from corners **70,210** and **290,120** of height **10**, the second of corners **120,200,10** and **240,120,10** and of height **80**.
2. Place the screen in the **ViewCube/Front** view and **Zoom** to **1**.
3. *Click* the **Wedge** tool icon in the **Home/3D Modeling** panel. The command line shows:

```
Command: _wedge
Specify first corner or [Center]: 120,170,10
Specify other corner or [Cube/Length]: 80,160,10
Specify height or [2Point]: 70
Command: right-click
WEDGE
```

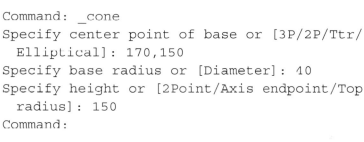

```
Specify first corner of wedge or [Center]:
  240,170,10
Specify corner or [Cube/Length]: 280,160,10
Specify height or [2Point]: 70
Command:
```

4. Place the screen in the **ViewCube/Isometric** view and **Zoom** to **1**.
5. Call the **Union** tool from the **Home/Solid Editing** panel and in response to the prompts in the tool's sequences *pick* each of the 4 objects in turn to form a union of the 4 objects.

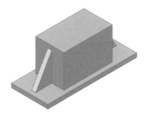

Fig. 12.29 Fourth example – **3D Objects**

The result is shown in Fig. 12.29.

Fifth example – **Cylinder** and **Torus** (Fig. 12.30)

1. Using the **Cylinder** tool from the **Home/3D Modeling** panel, construct a cylinder of centre **180,160**, of radius **40** and height **120**.
2. *Click* the **Torus** tool icon in the **Home/3D Modeling** panel. The command line shows:

```
Command: _torus
Specify center point or [3P/2P/Ttr]: 180,160,10
Specify radius or [Diameter]: 40
Specify tube radius or [2Point/Diameter]: 10
Command: right-click
TORUS
Specify center point or [3P/2P/Ttr]: 180,160,110
Specify radius or [Diameter] <40>: right-click
Specify tube radius or [2Point/Diameter] <10>:
  right-click
Command:
```

3. Call the **Cylinder** tool and construct another cylinder of centre **180,160**, of radius **35** and height **120**.
4. Place in the **ViewCube/Isometric** view and **Zoom** to **1**.
5. *Click* the **Union** tool icon in the **Home/Solid Editing** panel and form a union of the larger cylinder and the two tori.
6. *Click* the **Subtract** tool icon in the **Home/Solid Editing** panel and subtract the smaller cylinder from the union.

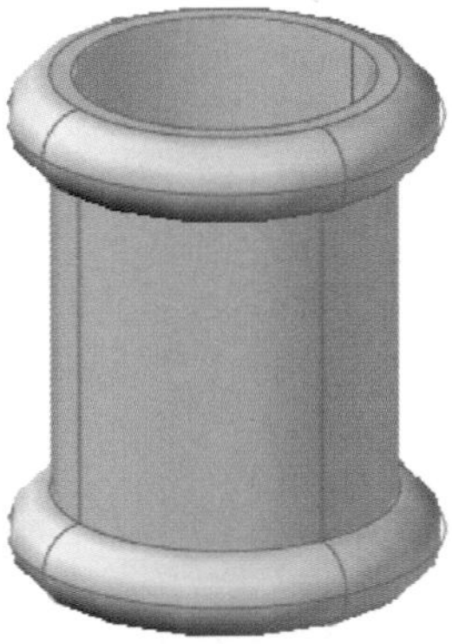

Fig. 12.30 Fifth example – **3D Objects**

The result is shown in Fig. 12.30.

The **Chamfer** and **Fillet** tools

The **Chamfer** and **Fillet** tools from the **Home/Modify** panel (Fig. 12.31), used to create chamfers and fillets in 2D drawings in AutoCAD 2009, can just as well be used when constructing 3D models.

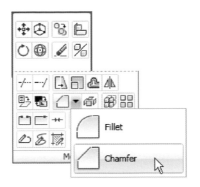

Fig. 12.31 The **Chamfer** and **Fillet** tools in the **Home**/**Modify** panel

Example – Chamfer and Fillet (Fig. 12.34)

1. Working to the sizes given in Fig. 12.32 and using the **Box** and **Cylinder** tools, construct the 3D model shown in Fig. 12.33.
2. Place in the **ViewCube**/**Isometric** view. **Union** the two boxes and with the **Subtract** tool, subtract the cylinders from the union.

Note

To construct the elliptical cylinder:

```
Command: _cylinder
Specify center point of base or [3P/2P/Ttr/
  Elliptical]: enter e right-click
Specify endpoint of first axis or [Center]:
  130,160
Specify other endpoint of first axis: 210,160
Specify endpoint of second axis: 170,180
Specify height or [2Point/Axis endpoint]: 50
Command:
```

---

3. *Click* the **Fillet** tool icon in the **Home**/**Modify** panel (Fig. 12.31). The command line shows:

```
Command: _fillet
Current settings: Mode = TRIM. Radius = 1
Specify first object or [Undo/Polyline/Radius/Trim/
  Multiple]: enter r (Radius) right-click
Specify fillet radius <1>: 10
Select first object: pick one corner
Select an edge or [Chain/Radius]: pick a second
  corner
```

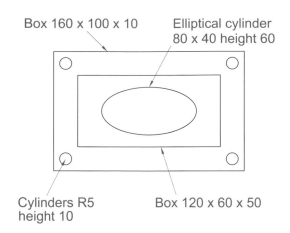

Box 160 x 100 x 10   Elliptical cylinder
                        80 x 40 height 60

Cylinders R5   Box 120 x 60 x 50
height 10

Fig. 12.32 Example – **Chamfer** and **Fillet** – sizes for the model

```
Select an edge or [Chain/Radius]: pick a third
  corner
Select an edge or [Chain/Radius]: pick the fourth
  corner
Select an edge or [Chain/Radius]: right-click
4 edge(s) selected for fillet.
Command:
```

Figure 12.33 shows the 3D model so far constructed.

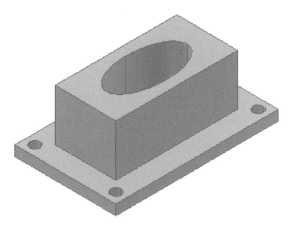

Fig. 12.33 Example – isometric view – **Chamfer** and **Fillet** – the model before using the tools

4. *Click* the **Chamfer** tool in the **Home/Modify** panel (Fig. 12.31). The command line shows:

```
Command: _chamfer
(TRIM mode) Current chamfer Dist1 = 1, Dist2 = 1
Select first line or [Undo/Polyline/Distance/
  Angle/Trim/method/Multiple] enter d (Distance)
  right-click
```

```
Specify first chamfer distance <1>: 10
Specify second chamfer distance <10>: 10
Select first line: pick one corner One side of the
  box highlights
Base surface selection … Enter surface selection
  [Next/OK (current)] <OK>: right click
Specify base surface chamfer distance <10>:
  right-click
Specify other surface chamfer distance <10>:
  right-click
Select an edge or [Loop]: pick the edge again
  Select an edge: pick the second edge
Select an edge [or Loop]: right-click
Command:
```

and two edges are chamfered. Repeat to chamfer the other two edges.

Figure 12.34 shows the completed 3D model.

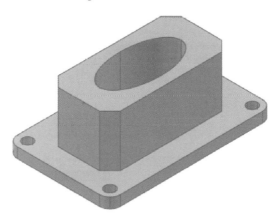

Fig. 12.34 Example – **Chamfer** and **Fillet**

Note on the tools Union, Subtract and Intersect

The tools **Union**, **Subtract** and **Intersect** found in the **Home/Solids Editing** panel are known as the **Boolean** operators after the mathematician George Boole. They can be used to form unions, subtractions or intersection between extrusions solids of revolution, or any of the 3D objects.

Note on using 2D **Draw** tools on 3D models

As was seen when using the **Move** tool from the **Home/Draw** panel and the **Chamfer** and **Fillet** tools from the **Home/Modify** panel, so can other tools – **Copy**, **Mirror**, **Rotate** and **Scale** from the **Home/Modify** panel – be used in connection with the construction of 3D models.

Constructing 3D surfaces using the **Extrude** tool

In this example of the construction of a 3D surface model the use of the **Dynamic Input** (DYN) method of construction will be shown.

1. Place the AutoCAD drawing area in the **View Block/Isometric** view.
2. *Click* the **Dynamic Input** button in the status bar to make dynamic input active.

Example – **Dynamic Input** (Fig. 12.36)

1. Using the line tool construct the outline shown in Fig. 12.35.
2. Call the **Extrude** tool and window the line outline.
3. Extrude to a height of **100**.

The stages of producing the extrusion are shown in Figs 12.35 and 12.36. The resulting 3D model is a surface model.

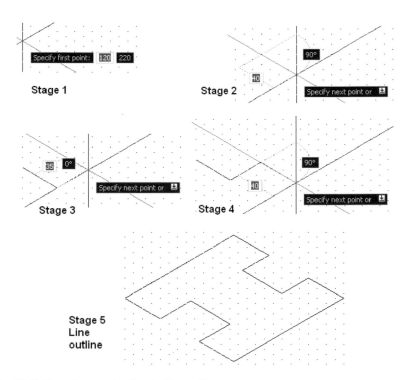

Stage 1

Stage 2

Stage 3

Stage 4

Stage 5
Line
outline

Fig. 12.35 Example – constructing the **Line** outline

The **Sweep** tool

To call the tool *click* on its tool icon in the **Home/3D Modeling** panel (Fig. 12.37).

Example – **Sweep** (Fig. 12.40)

1. Construct the pline outline (Fig. 12.38) in the **View Cube/Top** view.

CHAPTER 12

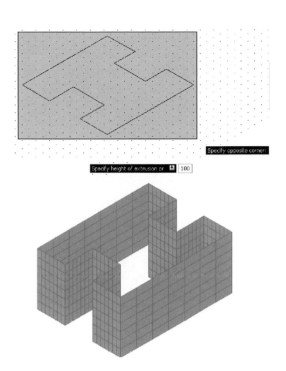

Fig. 12.36 Example – **Dynamic Input**

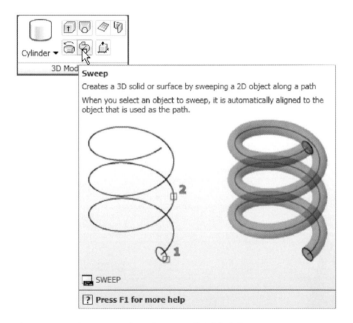

Fig. 12.37 Selecting the **Sweep** tool from the **Home/3D Modeling** panel

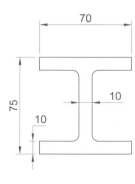

Fig. 12.38 Example –
Sweep – the outline to be
swept

2. Change to the **ViewCube/Front** view, **Zoom** to **1** and construct a pline as shown in Fig. 12.39 as a path central to the ellipse.
3. Place the window in a **View Cube/Isometric** view and *click* the **Sweep** tool icon (Fig. 12.37). The command line shows:

```
Command: _sweep
Current wire frame density: ISOLINES = 4
```

```
Select objects to sweep: pick the ellipse 1 found
Select objects to sweep: right-click
Select sweep path or [Alignment/Base point/Scale/
  Twist]: pick the pline
Command:
```

The result is shown in Fig. 12.40.

The **Loft** tool

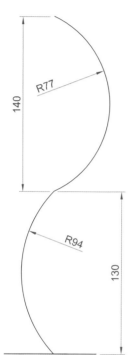

Fig. 12.39 Example –
Sweep – the pline path

To call the tool *click* on its icon in the **Home/3D Modeling** panel.

Example – **Loft** (Fig. 12.44)

1. Construct the seven circles shown in Fig. 12.41 at vertical distances of **30** units apart.
2. Place the drawing area in the **ViewCube/Isometric** view.
3. Call the **Loft** tool with a *click* on its tool icon in the **Home/3D Modeling** panel (Fig. 12.42).
4. The command line shows:

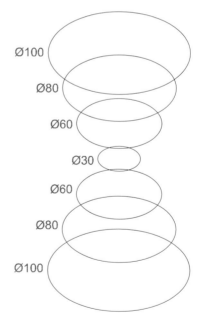

Fig. 12.41 Example – **Loft** – the cross-sections

```
Command: _loft
Select cross-sections in lofting order: <Snap
  off> pick the bottom circle 1 found
Select cross-sections in lofting order: pick the
  next circle 1 found, 2 total
Select cross-sections in lofting order: pick the
  next circle 1 found, 3 total
```

Fig. 12.40 Example –
Sweep

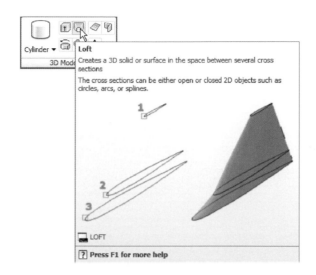

Fig. 12.42 Selecting the **Loft** tool from the **Home/3D Modeling** panel

```
Select cross-sections in lofting order: pick the
  next circle 1 found, 4 total
Select cross-sections in lofting order: pick the
  next circle 1 found, 5 total
Select cross-sections in lofting order: pick the
  next circle 1 found, 6 total
Select cross-sections in lofting order: pick the
  next circle 1 found, 7 total
Select cross-sections in lofting order: right-click
```

and the **Loft Settings** dialog appears (Fig. 12.43).

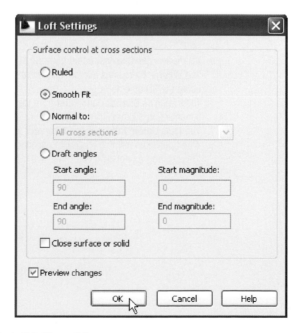

Fig. 12.43 The **Loft Settings** dialog

5. *Click* the **Smooth Fit** radio button to make it active, followed by a *click* on the **OK** button. The loft appears.

The result is shown in Fig. 12.44.

Fig. 12.44   Example – **Loft**

REVISION NOTES

1. In the AutoCAD 3D coordinate system, positive Z is towards the operator away from the monitor screen.
2. A 3D face is a mesh behind which other details can be hidden.
3. The **Extrude** tool can be used for extruding closed plines or regions to stated heights, to stated slopes or along paths.
4. The **Revolve** tool can be used for constructing solids of revolution through any angle up to 360°.
5. 3D models can be constructed from the 3D objects **Box**, **Sphere**, **Cylinder**, **Cone**, **Torus** and **Wedge**. Extrusions and/or solids of revolutions may form part of models constructed using the 3D objects.
6. Tools such as **Chamfer** and **Fillet** from the **Home/Modify** panel can be used when constructing 3D models.
7. The tools **Union**, **Subtract** and **Intersect** are known as the **Boolean** operators.
8. When polylines forming an outline which is not closed are acted upon by the **Extrude** tool the resulting models will be 3D Surface models.

Exercises

Methods of constructing answers to the following exercises can be found in the free website:

http://books.elsevier.com/companions/9780750689830

The exercises which follow require the use of tools from the **Home/3D Modeling** panel in association with tools from other panels.

1. Figure 12.45 shows the pline outline from which the polysolid outline of Fig. 12.46 has been constructed to a height of **100** and **Width** of **3**. When the polysolid has been constructed, construct extrusions can then be subtracted from the polysolid. Sizes of the extrusions are left to your judgement.

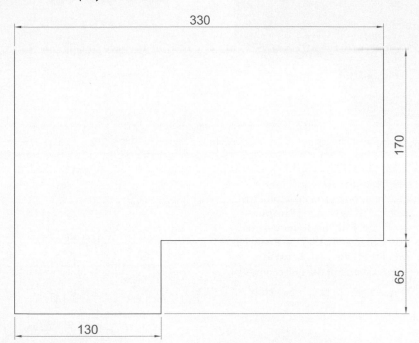

Fig. 12.45 Exercise 1 – outline for polyline

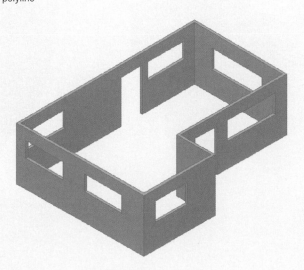

Fig. 12.46 Exercise 1

2. Figure 12.47 shows a 3D model constructed from four polysolids which have been formed into a union using the **Union** tool from the **Modify** panel. The original polysolid was formed from a hexagon of edge length **30**. The original polysolid was of height **40** and **Width 5**. Construct the union.

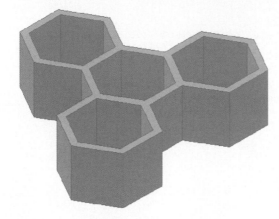

Fig. 12.47 Exercise 2

3. Figure 12.48 shows the 3D model from Exercise 2 acted upon by the **Presspull** tool, which can be called by *entering* **Prespull** at the command line. With the 3D model from Exercise 2 on screen and using the **Presspull** tool, construct the 3D model shown in Fig. 12.46. The distance of the pull can be estimated.

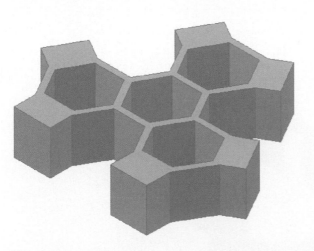

Fig. 12.48 Exercise 3

4. Construct the 3D model of a wine glass as shown in Fig. 12.50, working to the dimensions given in the outline drawing of Fig. 12.49. You will need to construct the outline and change it into a region before being able to change the outline into a solid of revolution using the **Revolve** tool from the **Home/3D Modeling** panel. This is because the semi-elliptical part of the outline has been constructed using the **Ellipse** tool, resulting in part of the outline being a spline, which cannot be acted upon by **Polyline Edit** to form a closed pline.

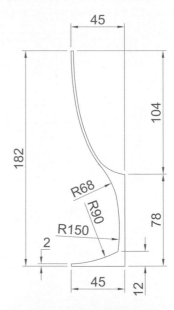

Fig. 12.49 Exercise 4 – outline drawing

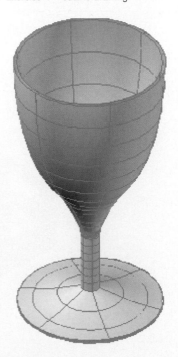

Fig. 12.50 Exercise 4

5. Figure 12.51 shows the outline from which a solid of revolution can be constructed. Use the **Revolve** tool from the **Home/3D Modeling** panel to construct the solid of revolution.

6. Construct a 3D solid model of a bracket working to the information given in Fig. 12.52.

7. Working to the dimensions given in Fig. 12.53 construct an extrusion of the plate to a height of **5** units.

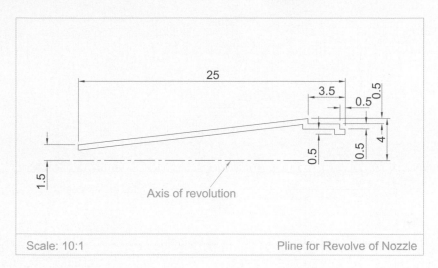

Fig. 12.51 Exercise 5

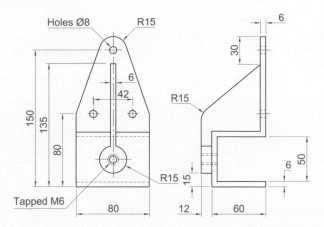

Fig. 12.52 Exercise 6

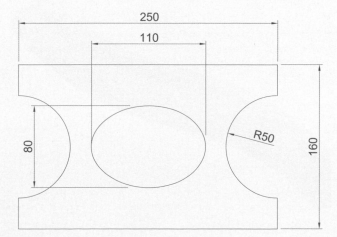

Fig. 12.53 Exercise 7

CHAPTER 12

8. Working to the details given in the orthographic projection of Fig. 12.54, construct a 3D model of the assembly. After constructing the pline outline(s) required for the solid(s) of revolution, use the **Revolve** tool to form the 3D solid.

9. Working to the polylines shown in Fig. 12.55 construct the **Sweep** shown in Fig. 12.56.

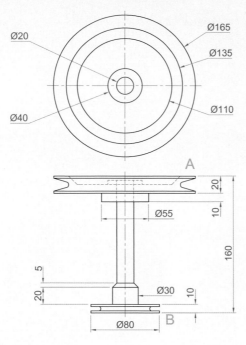

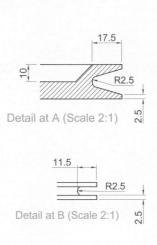

Detail at A (Scale 2:1)

Detail at B (Scale 2:1)

Fig. 12.54 Exercise 8

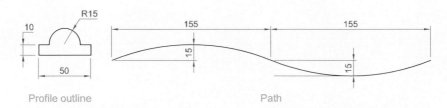

Profile outline

Path

Fig. 12.55 Exercise 9 – profile and path dimensions

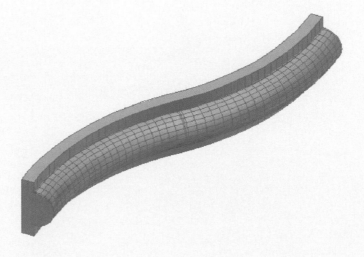

Fig. 12.56 Exercise 9

10. Construct the cross-sections as shown in Fig.
    12.57 working to suitable dimensions. From
    the cross-sections construct the lofts shown in
    Fig. 12.58. The lofts are topped with a sphere
    constructed using the **Sphere** tool.

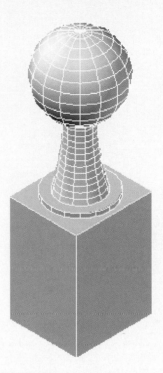

Fig. 12.58 Exercise 10

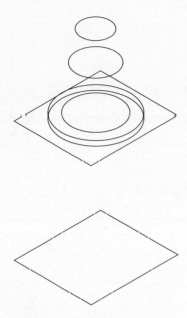

Fig. 12.57 The cross-sections for Exercise 10

Chapter 13

3D models in viewports

AIM OF THIS CHAPTER

The aim of this chapter is to give examples of 3D solid models constructed in multiple viewport settings.

Setting up viewport systems

One of the better methods of constructing 3D models is in different multiple viewports. This allows what is being constructed to be seen from a variety of viewing positions. Set up multiple viewports as follows:

1. *Click* **New** in the **View/Viewports** panel (Fig. 13.1). The **Viewports** dialog appears (Fig. 13.2). In the dialog:

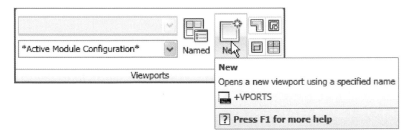

Fig. 13.1 Selecting **New** from the **View/Viewports** panel.

2. *Click* the **New Viewports** tab and a number of named viewports systems appears in the **Standard Viewports** list in the dialog.
3. *Click* the name **Four: Equal**, followed by a *click* on **3D** in the **Setup** pop-up list. A preview of the **Four: Equal** viewports screen appears showing the views appearing in each of the four viewports.
4. *Click* in each viewport in the dialog in turn, followed by selecting **Conceptual** from the **Visual Style** pop-up list.
5. *Click* the **OK** button of the dialog and the AutoCAD 2009 drawing area appears showing the four viewport layout.
6. *Click* in each viewport in turn and **Zoom** to **All**.

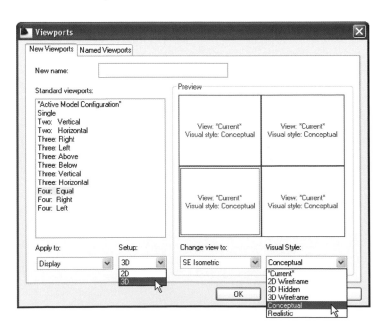

Fig. 13.2 The **Viewports** dialog

The result is shown in Fig. 13.3.

First example – **Four: Equal** viewports (Fig. 13.7)

Figure 13.4 shows a first angle orthographic projection of a support. To construct a **Scale 1:1** 3D model of the support in a **Four: Equal** viewport setting, proceed as follows:

1. *Click* **View** in the **View/Viewports** panel. In the **Viewports** dialog make sure the **3D** option is selected from the **Setup** pop-up list and **Conceptual** from the **Visual Style** pop-up menu and *click* the **OK** button of the dialog. The AutoCAD 2009 drawing area appears in a **Four: Equal** viewport setting.
2. *Click* in each viewport in turn, making the selected viewport active, and **Zoom** to **1**.
3. Using the **Polyline** tool, construct the outline of the plan view of the plate of the support, including the holes in the **Top** viewport (Fig. 13.5). Note the views in the other viewports.
4. Call the **Extrude** tool from the **Home/3D Modeling** panel and extrude the plan outline and the circles to a height of **20**.
5. With **Subtract** from the **Home/Solids Editing** panel, subtract the holes from the plate (Fig. 13.6).
6. Call the **Box** tool and in the centre of the plate construct a box of Width = **60**, Length = **60** and Height = **30**.
7. Call the **Cylinder** tool and in the centre of the box construct a cylinder of Radius = **20** and of Height = **30**.

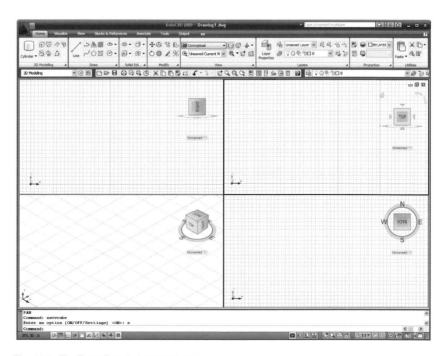

Fig. 13.3 The **Four: Equal** viewports layout

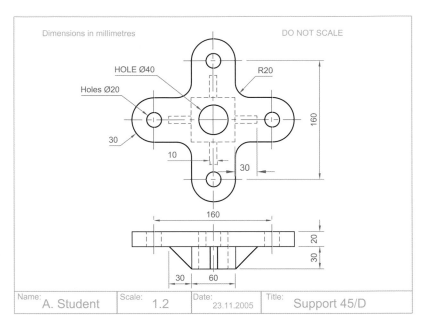

Fig. 13.4 Orthographic projection of the support for the first example

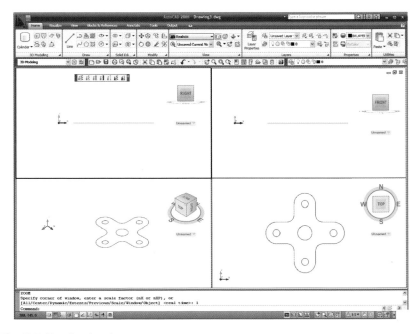

Fig. 13.5 The plan view drawn

8. Call **Subtract** and subtract the cylinder from the box.
9. *Click* in the **Right** viewport, and, with the **Move** tool, move the box and its hole into the correct position with regard to the plate.
10. With **Union**, form a union of the plate and box.
11. *Click* in the **Front** viewport and construct a triangle of one of the webs attached between the plate and the box. With **Extrude**, extrude the triangle to a height of **10**. With the **Mirror** tool, mirror the web to the other side of the box.

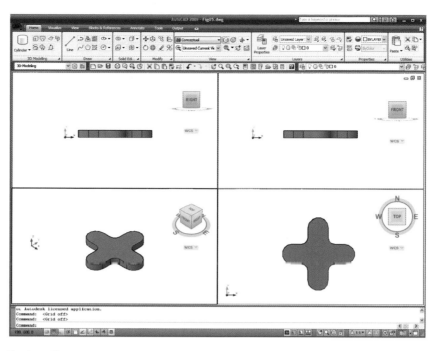

Fig. 13.6 The four views after using the **Extrude** and **Subtract** tools

12. *Click* in the **Right** viewport and with the **Move** tool, move the two webs into their correct position between the box and plate. Then, with **Union**, form a union between the webs and the 3D model.

13. In the **Right** viewport, construct the other two webs and in the **Front** viewport, move, mirror and union the webs as in steps **12** and **13**.

Figure 13.7 shows the resulting 4-viewport scene.

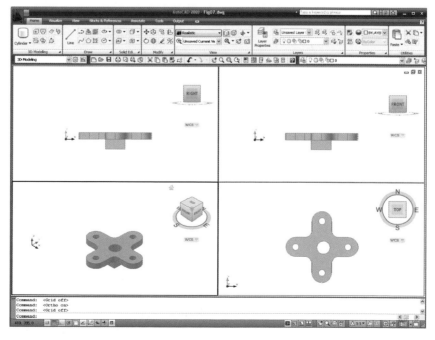

Fig. 13.7 First example – **Four: Equal** viewports

Second example – **Four: Left** viewports (Fig. 13.9)

1. Open the **Four: Left** viewport layout from the **Viewports** dialog.
2. Make a new layer of colour magenta and make that layer current.
3. In the **Top** viewport construct an outline of the web of the Support Bracket shown in Fig. 13.8. With the **Extrude** tool, extrude the parts of the web to a height of **20**.

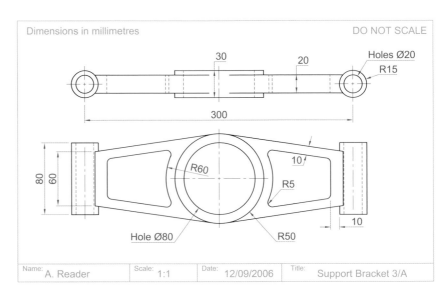

Fig. 13.8 Working drawing for the second example

4. With the **Subtract** tool, subtract the holes from the web.
5. In the **Top** viewport, construct two cylinders central to the extrusion, one of radius **50** and height **30**, the second of radius **40** and height **30**. With the **Subtract** tool, subtract the smaller cylinder from the larger.
6. *Click* in the **Front** viewport and move the cylinders vertically by **5** units. With **Union** form a union between the cylinders and the web.
7. Still in the **Front** viewport and at one end of the union, construct two cylinders, the first of radius **10** and height **80**, the second of radius **15** and height **80**. Subtract the smaller from the larger.
8. With the **Mirror** tool, mirror the cylinders to the other end of the union.
9. Make the **Top** viewport current and with the **Move** tool, move the cylinders to their correct position at the ends of the union. Form a union between all parts on screen.
10. Make the **Isometric** viewport current. From the **Home/View** panel select **Realistic**.

Figure 13.9 shows the result.

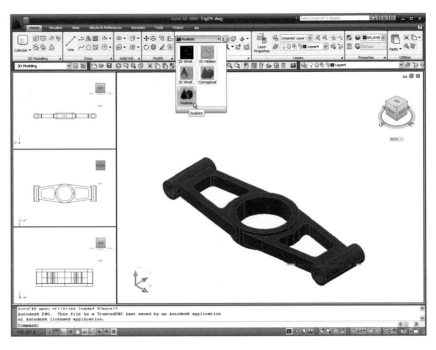

Fig. 13.9 Second example – **Four: Left** viewports

Third example – **Three: Right** viewports (Fig. 13.11)

1. Open the **Three: Right** viewport layout from the **Viewports** dialog. Make sure the **3D** setup is chosen.
2. Make a new layer of colour **Green** and make that layer current.
3. In the **Front** viewport (top left), construct a pline outline to the dimensions in Fig. 13.10.
4. Call the **Revolve** tool from the **Home/3D Modeling** panel and revolve the outline through 360 °.
5. In each of the three viewports in the **Home/View** panel select **Conceptual** from its pop-up list.

The result is shown in Fig. 13.11.

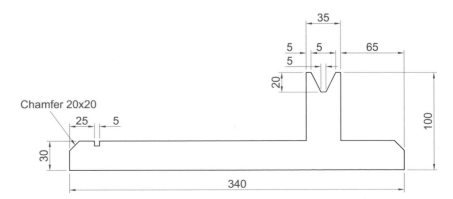

Fig. 13.10 Third example – outline for solid of revolution

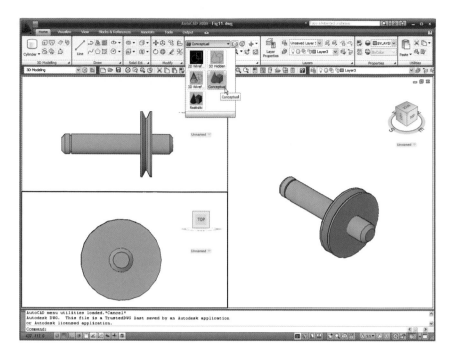

Fig. 13.11 Third example – **Three: Right** viewports

Note

1. When working in viewport layouts such as in the above three examples, it is important to make good use of the **Zoom** tool, mainly because the viewports are smaller than the single viewport when working in AutoCAD 2009.

2. As in all other forms of constructing drawings in AutoCAD 2009 frequent toggling of **SNAP**, **ORTHO** and **GRID** will allow speedier and more accurate working.

REVISION NOTES

1. Outlines suitable for use when constructing 3D models can be constructed using the 2D tools such as **Line**, **Arc**, **Circle**, and **Polyline**. Such outlines must be changed either to closed polylines or to regions before being incorporated in 3D models.
2. The use of multiple viewports can be of value when constructing 3D models in that various views of the model appear, enabling the operator to check the accuracy of the 3D appearance throughout the construction period.

Exercises

Methods of constructing answers to the following exercises can be found in the free website:

http://www.books.elsevier.com/companions/9780750689830

1.   Using the **Cylinder**, **Box**, **Sphere**, **Wedge** and **Fillet** tools, together with the **Union** and **Subtract** tools and working to any sizes thought suitable, construct the 'head' as shown in the **Three: Right** viewport as shown in Fig. 13.12.

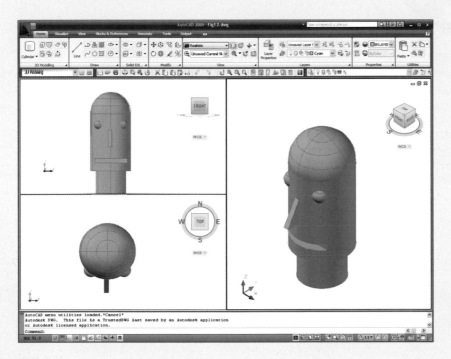

Fig. **13.12**   Exercise 1

2.   Using the tools **Sphere**, **Box**, **Union** and **Subtract** and working to the dimensions given in Fig. 13.14, construct the 3D solid model as shown by the isometric drawing in Fig. 13.13.

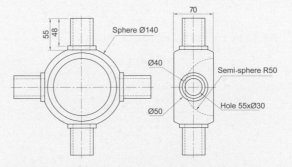

Fig. **13.14**   Exercise 2 – working drawings

Fig. **13.13**   Exercise 2

3.   Each link of the chain shown in Fig. 13.15 has been constructed using the tool **Extrude** and extruding a small circle along an elliptical path. Copies of the link were then made, half of which were rotated in a **Right** view and then

moved into their position relative to the other links. Working to suitable sizes construct a link and from the link construct the chain as shown.

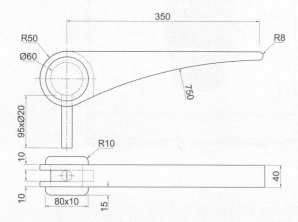

Fig. 13.15 Exercise 3

4.   A two-view orthographic projection of a rotatable lever from a machine is given in Fig. 13.16, together with an isometric drawing of the 3D model constructed to the details given in the drawing of Fig. 13.17. Construct the 3D model drawing in a **Four: Equal** viewport setting.

Fig. 13.16 Exercise 4 – orthographic projection

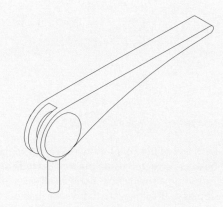

Fig. 13.17 Exercise 4

5.   Working in a **Three: Left** viewport setting, construct a 3D model of the faceplate to the dimensions given in Fig. 13.18. With the **Mirror** tool, mirror the model to obtain an opposite facing model. In the **Isometric** viewport call the **Hide** tool (Fig. 13.19).

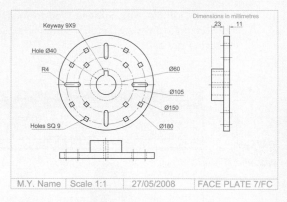

Fig. 13.18 Exercise 5 – dimensions

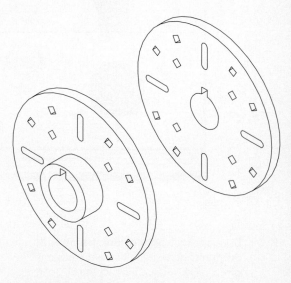

Fig. 13.19 Exercise 5

Chapter 14

The modification of 3D models

AIMS OF THIS CHAPTER

The aims of this chapter are:

1. to demonstrate how 3D models can be saved as blocks for insertion into other drawings via the **DesignCenter**;
2. to show how a library of 3D models in the form of blocks can be constructed to enable the models to be inserted into other drawings;
3. to give examples of the use of the tools from the **Home/Solid Editing** panel:
   - **3D Array** – **Rectangular** and **Polar 3D** arrays
   - **3D Mirror**
   - **3D Rotate**
4. to give examples of the use of the **Section** tool from the **Home/Solid Editing** panel;
5. to give examples of the use of the **Helix** tool;
6. to give an example of construction involving **Dynamic Input**;
7. to show how to obtain different views of 3D models in 3D space using the **ViewCube**.

Creating 3D model libraries

In the same way as 2D drawings of parts such as electronics symbols, engineering parts, building symbols and the like can be saved in a file as blocks and then opened into another drawing by *dragging* the appropriate block drawing from the **DesignCenter**, so can 3D models.

First example – inserting 3D blocks (Fig. 14.4)

1. Construct 3D models of the parts for a lathe milling wheel holder to details as given in Fig. 14.1, each on a layer of different colours.
2. Save each of the 3D models of the parts to file names as given in Fig. 14.1 as blocks using **Create** from the **Blocks & Reference/Block** panel. Save all seven blocks and delete the drawings on screen. Save the drawing with its blocks to a suitable file name (**Fig01.dwg**).

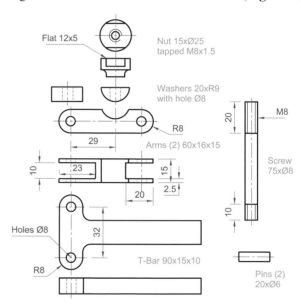

Fig. 14.1 The components of a lathe milling wheel holder

3. Set up a **Four: Equal** viewports setting.
4. Open the **DesignCenter** with a *click* on its icon in the **Standard** toolbar (Fig. 14.2), or by pressing the **Ctrl** and **2** keys of the keyboard.

Fig. 14.2 Calling the **DesignCenter** from the **Standard** toolbar

5. In the **DesignCenter** *click* the directory **Chap14**, followed by another *click* on **Fig01.dwg** and yet another *click* on **Blocks**. The saved blocks appear as icons in the right-hand area of the **DesignCenter**.

6. *Drag and drop* the blocks one by one into one of the viewports on screen. Figure 14.3 shows the **Nut** block ready to be *dragged* into the **Right** viewport. As the blocks are *dropped* on screen, they will need

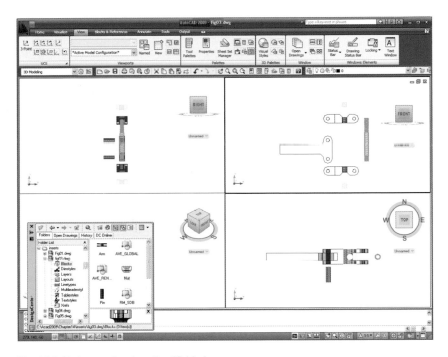

Fig. 14.3 First example – inserting 3D blocks

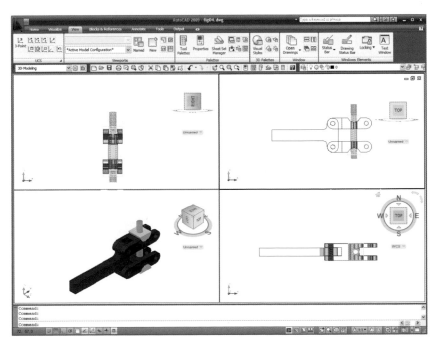

Fig. 14.4 First example – inserting 3D blocks

moving into their correct positions in suitable viewports using the **Move** tool from the **Home/Modify** panel.

7. Using the **Move** tool, move the individual 3D models into their final places on screen and shade the **Isometric** viewport using **Conceptual** shading from the the **Home/View** panel (Fig. 14.4).

Note

1. It does not matter which of the four viewports any one of the blocks is *dragged* and *dropped* into. The part automatically assumes the view of the viewport.

2. If a block destined for layer **0** is *dragged* and *dropped* into the layer **Center** (which in our **acadiso.dwt** is of colour **red** and of linetype **CENTER2**), the block will take on the colour (red) and linetype of that layer (**CENTER2**).

3. In this example, the blocks are 3D models and there is no need to use the **Explode** tool option.

Fig. 14.5 Second example – the five fastenings

Second example – a library of fastenings (Fig. 14.6)

1. Construct 3D models of a number of engineering fastenings. The number constructed does not matter. In this example only five have been constructed – a 10 mm round head rivet, a 20 mm countersunk head rivet, a cheese head bolt, a countersunk head bolt and a hexagonal head bolt together with its nut (Fig. 14.5). With the **Create** tool save each separately as a block, erase the original drawings and save the file to a suitable file name – in this example **Fig05.dwg**.

2. Open the **DesignCenter**, *click* on the **Chapter 14** directory, followed by a *click* on **Fig05.dwg**. Then *click* again on **Blocks** in the content list of **Fig05.dwg**. The five 3D models of fastenings appear as icons in the right-hand side of the **DesignCenter** (Fig. 14.6).

3. Such blocks of 3D models can be *dragged* and *dropped* into position in any engineering drawing where the fastenings are to be included.

Constructing a 3D model (Fig. 14.9)

A three-view projection of a pressure head is shown in Fig. 14.7. To construct a 3D model of the head, proceed as follows:

1. Set the **ViewCube/Front** view.

2. Construct the outline to be formed into a solid of revolution (Fig. 14.8) on a layer colour magenta, and, with the **Revolve** tool, produce the 3D model of the outline.

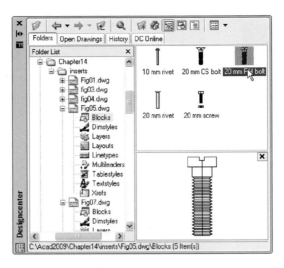

Fig. 14.6 Second example – a library of fastenings

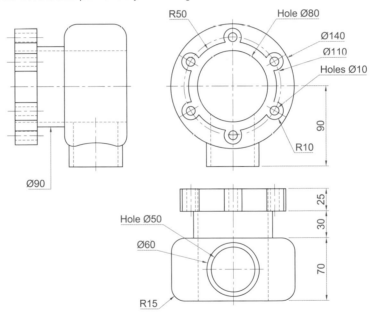

Fig. 14.7 Orthographic drawing for the example of constructing a 3D model

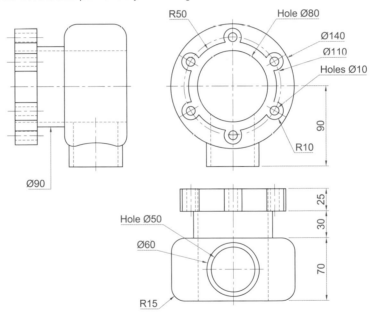

Fig. 14.8 Example of constructing as 3D model – outline for solid of revolution

3. Set the **ViewCube/Top** view and with the **Cylinder** tool, construct cylinders as follows:
   - in the centre of the solid – radius **50** and height **50**
   - with the same centre – radius **40** and height **40**. Subtract this cylinder from that of radius **50**
   - at the correct centre – radius **10** and height **25**
   - at the same centre – radius **5** and height **25**. Subtract this cylinder from that of radius **10**
4. With the **Array** tool, form a polar 6 times array of the last two cylinders based on the centre of the 3D model.
5. Set the **ViewCube/Front** view.
6. With the **Move** tool, move the array and the other two cylinders to their correct positions relative to the solid of revolution so far formed.

Fig. 14.9 Example of constructing a 3D model

7. With the **Union** tool form a union of the array and other two solids.
8. Set the **ViewCube/Right** view.
9. Construct a cylinder of radius **30** and height **25** and another of radius **25** and height **60** central to the lower part of the 3D solid so far formed.
10. Set the **ViewCube/Top** view and with the **Move** tool move the two cylinders into their correct position.
11. With **Union**, form a union between the radius **30** cylinder and the 3D model and with **Subtract**, subtract the radius **25** cylinder from the 3D model.
12. *Click* **Conceptual** in the **Home/View** panel list.

The result is given in Fig. 14.9.

> **Note**
>
> This 3D model could have been constructed equally as well in a three or four viewports setting.

The **3D Array** tool

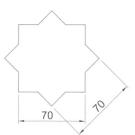

Fig. 14.10 Example – **3D Array** – the star pline

First example – a **Rectangular Array** (Fig. 14.12)

1. Construct the star-shaped pline on a layer colour green (Fig. 14.10) and extrude it to a height of **20**.

Fig. 14.11 Selecting **3D Array** from the **Modify/3D Operations** drop-down menu

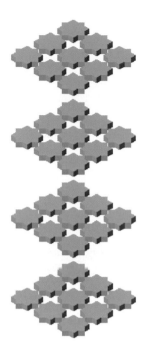

2. *Click* on the **3D Array** icon in the **Modify/3D Operations** menu (Fig. 14.11). The command line shows:

```
Command: _3darray
Select objects: pick the extrusion 1 found
Select objects: right-click
Enter the type of array [Rectangular/Polar]
  <R>: right-click
Enter the number of rows (—) <1>: enter 3
  right-click
Enter the number of columns (III): enter 3
  right-click
Enter the number of levels (...): enter 4
  right-click
Specify the distance between rows (—): enter 100
  right-click
Specify the distance between columns (III): enter
  100 right-click
Specify the distance between levels (...): enter
  300 right-click
Command:
```

Fig. 14.12 First example – a **Rectangular Array**

3. Place the screen in the **ViewCube/Isometric** view.
4. Shade using the **Home/View/Conceptual** visual style (Fig. 14.12).

Second example – a **Polar Array** (Fig. 14.13)

1. Use the same star-shaped 3D model.
2. Call the **3D Array** tool again. The command line shows:

```
Command: _3darray
Select objects: pick the extrusion 1 found
Select objects: right-click
```

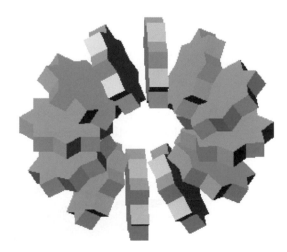

Fig. 14.13 Second example – a **Polar Array**

```
Enter the type of array [Rectangular/Polar]
  <R>: enter p (Polar) right-click
Enter number of items in the array: 12
Specify the angle to fill (+=ccw), -=cw)
  <360>: right-click
Rotate arrayed objects? [Yes/No] <Y>:
  right-click
Specify center point of array: 235,125
Specify second point on axis of rotation:
  300,200
Command:
```

3. Place the screen in the **ViewCube/Isometric** view.
4. Shade using the **Home/View/Conceptual** visual style (Fig. 14.13).

Third example – a **Polar Array** (Fig. 14.15)

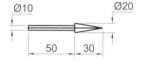

Fig. 14.14 Third example –
a **Polar Array** – the 3D
model to be arrayed

1. Working on a layer of colour red, construct a solid of revolution in the form of an arrow to the dimensions as shown in Fig. 14.14.
2. *Click* **3D Array** from the **Modify/3D Operations** drop-down menu. The command line shows:

```
Command: _3darray
Select objects: pick the arrow 1 found
Select objects: right-click
Enter the type of array [Rectangular/Polar]
  <R>: enter p right-click
Enter the number of items in the array: enter 12
  right-click
Specify the angle to fill (+=ccw, -=cw)
  <360>: right-click
Rotate arrayed objects? [Yes/No] <Y>:
  right-click
Specify center point of array: enter 40,170,20
  right-click
Specify second point on axis of rotation: enter
  60,200,100 right-click
Command:
```

3. Place the array in the **ViewCube/Isometric** view and shade to **Visual Styles/Realistic**. The result is shown in Fig. 14.15.

Fig. 14.15 Third example – a **Polar Array**

The **Mirror** 3D tool

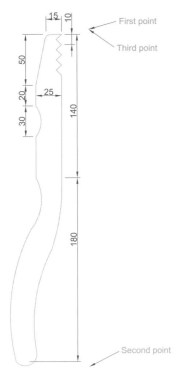

First point
Third point

Second point

Fig. 14.16 First example – **3D Mirror** – outline of object to be mirrored

First example – **3D Mirror** (Fig. 14.18)

1. Working on a layer colour green, construct the outline in Fig. 14.16.
2. Extrude the outline to a height of **20**.
3. Extrude the region to a height of **5** and render. A **Conceptual** style shading is shown in Fig. 14.17 (left-hand drawing).
4. *Click* on **3D Mirror** in the **3D Operation** sub-menu of the **Modify** drop-down menu. The command line shows:

```
Command: _3dmirror
Select objects: pick the extrusion 1 found
Select objects: right-click
Specify first point of mirror plane
  (3 points): pick
Specify second point on mirror plane: pick
Specify third point on mirror plane or
[Object/Last/Zaxis/View/XY/YZ/ZX/3points]:
enter .xy right-click
of (need Z): enter 1 right-click
Delete source objects? [Yes/No]: <N>:
  right-click
Command:
```

The result is shown in the right-hand illustration of Fig. 14.17.

Second example – **3D Mirror** (Fig. 14.19)

1. Construct a solid of revolution in the shape of a bowl in the **ViewCube/ Front** view working on a layer of colour yellow (Fig. 14.18).
2. *Click* **3D Mirror** in the **Modify/3D Operations** drop-down menu. The command line shows:

```
Command: _3dmirror
Select objects: pick the bowl 1 found
Select objects: right-click
Specify first point on mirror plane (3 points):
  pick
Specify second point on mirror plane: pick
Specify third point on mirror plane: enter .xy
  right-click
(need Z): enter 1 right-click
Delete source objects:? [Yes/No]: <N>:
  right-click
Command:
```

The result is shown in Fig. 14.19.

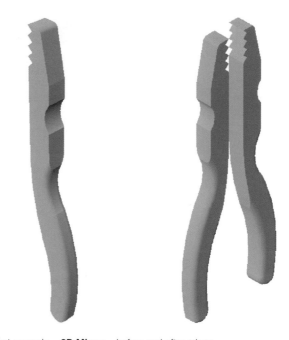

Fig. 14.18 Second example – **3D Mirror** – the 3D model

Fig. 14.17 First example – **3D Mirror** – before and after mirror

3. Place in the **ViewCube/Isometric** view.
4. Shade using the **Home/View/Conceptual** visual style (Fig. 14.19).

The **3D Rotate** tool

Example – **3D Rotate** (Fig. 14.20)

1. Use the same 3D model of a bowl as for the last example. Call the **3D Rotate** tool from the **3D Operations** sub-menu of the **Modify** drop-down menu. The command line shows:

```
Command: pick 3D Rotate from the Modify drop-down
   menu
3DROTATE
```

Fig. 14.19 Second example – **3D Mirror** – the result in a front view

```
Current positive angle in UCS: ANGDIR=
   counterclockwise ANGBASE=0
Select objects: pick the bowl 1 found
Select objects: right-click
Specify base point: pick the center bottom of the
   bowl
Specify rotation angle or [Copy/Reference] <0>:
   enter 60 right-click
Command
```

2. Place in the **ViewCube/Isometric** view and in **Conceptual** shading.

The result is shown in Fig. 14.20.

Fig. 14.20 Example – **3D Rotate**

The **Slice** tool

First example – **Slice** (Fig. 14. 24)

1. Construct a 3D model of the rod link device shown in the two-view projection in Fig. 14.21 on a layer colour green.
2. Place the 3D model in the **ViewCube/Top** view.
3. *Call* the **Slice** tool from the **Home/Solid Editing** panel (Fig. 14.22).

The command line shows:

```
Command: _slice
Select objects: pick the 3D model
Select objects to slice: right-click
Specify start point of slicing plane or [planar
   Object/Surface/Zaxis/View/XY/YZ/ZX/3points]
   <3points>: pick
```

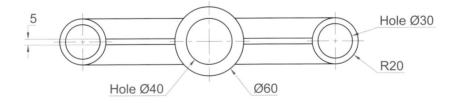

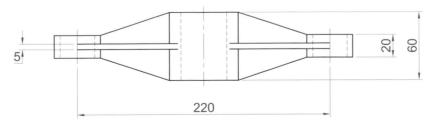

Fig. **14.21** First example – **Slice** – the two-view drawing

Fig. 14.22 The **Slice** tool icon from the **Home/Solid Editing** panel

```
Specify second point on plane: pick
Specify a point on desired side or [keep Both
  sides]<Both>: right-click
Command:
```

Figure 14.23 shows the *picked* points.

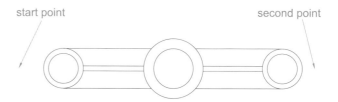

start point   second point

Fig. 14.23 First example – **Slice** – the *pick* points

4. With the **Move** tool, move the lower half of the sliced model away from the upper half.
5. Place the 3D model(s) in the **ViewCube/Isometric** view.
6. Shade in **Conceptual** visual style. The result is shown in Fig. 14.24.

Second example – **Slice** (Fig. 14.25)

Fig. 14.24 First example – **Slice**

1. Construct the closed pline (left-hand drawing in Fig. 14.25) and with the **Revolve** tool, form a solid of revolution from the pline
2. With the **Slice** tool and working to the same sequence as for the first **Slice** example, form two halves of the 3D model and render.

The right-hand illustration in Fig. 14.25 shows the result.

The **Section tool**

First example – **Section** (Fig. 14.28)

1. Construct a 3D model to the information given in Fig. 14.27 on layers of different colours. Note there are three objects in the model – a box, a lid and a cap.

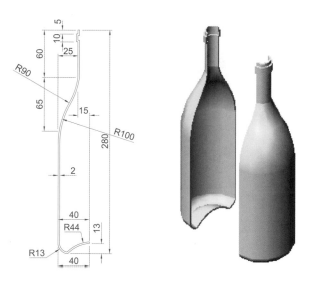

Fig. 14.25 Second example – **Slice**

2. Place the model in the **ViewCube/Top** view.
3. Make a new layer **Hatch** of colour **Red.**
4. *Click* the **Section Plane** tool icon in the **Home/3D Modeling** panel (Fig. 14.26). The command line shows:

Fig. 14.26 The **Section Plane** tool in the **Home/Solid Editing** panel

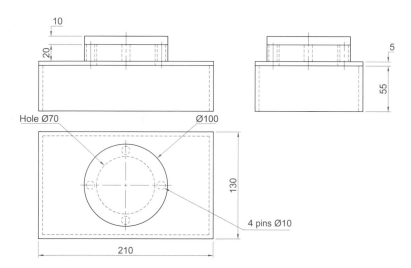

Fig. 14.27 First example – **Section** – orthographic projection

```
Command: _section
Select objects: window the model 3 found
Select objects: right-click
Select first point on Section plane by [Object/
  Zaxis/View/XY/YZ/ZX/3points <3points>: pick
  first point
Specify second point on plane: pick first point
Specify third point on plane: enter .xy
  right-click
of (need Z): enter 1 right-click
Command:
```

5. Place the drawing in the **ViewCube/Front** view.
6. Close all layers except **Hatch**.
7. Shade in realistic mode.

The result is shown in Fig. 14.28.

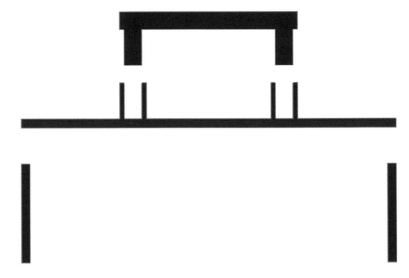

Fig. 14.28 First example – **Section**

Second example – **Section** (Fig. 14.29)

1. Open the drawing of the lathe tool holder constructed in answer to the first example in this chapter (Fig. 14.4). The drawing is in a **Four: Equal** viewports setting. In the **Top** viewport and from the **View** drop-down menu *click* **1 Viewport** in the **Viewports** sub-menu. The assembly appears in a full size single viewport.
2. Call the **Section** tool and proceed as in the First example – **Section** above making the section on a layer **Hatch** of colour **Cyan** and shade in **Conceptual** mode. The result is shown in Fig. 14.29.

Fig. 14.29 Second example – **Section**

Views of 3D models

World-Space ▽

Fig. 14.30 The **ViewCube** in its isometric posiiton

Figure 14.31 shows a two-view projection of a model of an arrow.

Some of the possible viewing positions of a 3D model which can be obtained by using the **ViewCube** have already been seen in this book. Figure 14.32 shows the viewing positions of the 3D model of the arrow using the viewing positions from the **ViewCube**. A *click* on the house icon at the top left-hand of the **ViewCube** presents an isometric view (Fig. 14.30).

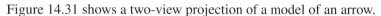
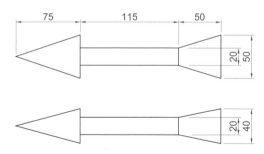

Fig. 14.31 Two views of the arrow

The **Viewpoint Presets** dialog

Other methods of obtaining a variety of viewing positions of a 3D model are: the **UCS**, described in a later chapter; using the **Viewport Presets** dialog called with a *click* on **Viewpoint Presets...** in the **View/3D Views** drop-down menu (Fig. 14.33); selection from the pop-up menu in the **Home/View** panel (Fig. 14.33).

The **Viewpoint Presets** dialog appears with a 3D model on screen, *entering* figures for degrees in the **X Axis** and **XY Plane** fields, followed by a *click* on the dialog's **OK** button causing the model to take up the viewing position indicated by these two angles.

CHAPTER 14

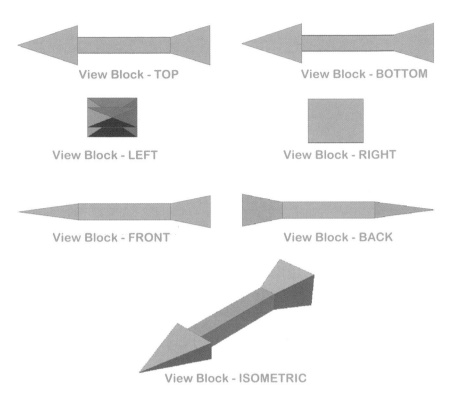

View Block - TOP

View Block - BOTTOM

View Block - LEFT

View Block - RIGHT

View Block - FRONT

View Block - BACK

View Block - ISOMETRIC

Fig. 14.32 The views using the **ViewCube**

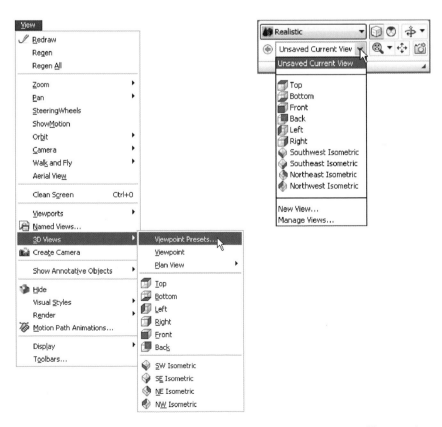

Fig. 14.33 Calling the **Viewpoint Presets** dialog and the pop-up menu in the **Home/View panel**

Note

The **Relative to UCS** radio button must be checked on to allow the 3D model to position along the two angles.

Example – **Viewpoint Presets**

1. With the 3D model of the arrow on screen, *click* **Viewpoint Presets...** in the **3D Views** sub-menu of the **Views** drop-down menu. The dialog appears.
2. *Enter* **330.0** in the **From X Axis** and **−30.0** in the **From XY Plane** fields and *click* the **OK** button of the dialog.
3. The 3D model takes up the viewing position indicated by the two angles (Fig. 14.34).

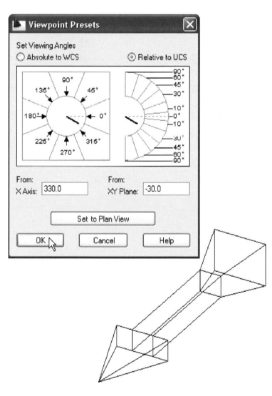

Fig. 14.34 Example – **Viewpoint Presets**

The **Helix** tool

The **Helix** tool can be called with a *click* on its tool icon in the extension of the **Home/3D Modeling** panel (Fig. 14.35).

First example – **Helix** (Fig. 14.37)

1. Construct the triangular outline shown in Fig. 14.34 using the **Polyline** tool. Make sure the pline outline is placed at right angles to the bottom

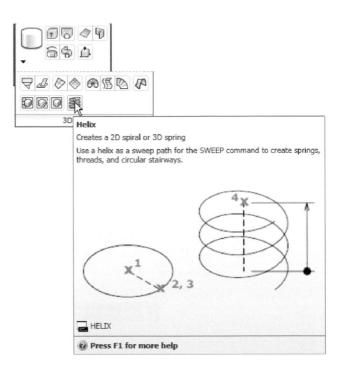

Fig. 14.35 The **Helix** tool icon in the extension of the **Home**/**3D Modeling** panel

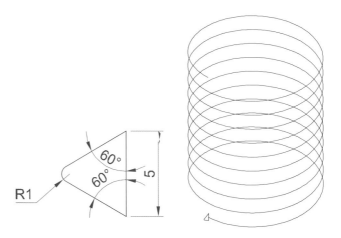

Fig. 14.36 First example – **Helix** – the polyline outline and the helix

end of the helix as shown in Fig. 14.35. This may mean moving and rotating the outline in a selection of the **ViewCube** views.

2. Make a new layer of colour **Cyan** and make it current.
3. Call the **Helix** tool from the **Home**/**3D Modeling** panel (Fig. 14.35) or from the **Modeling** toolbar. The command line shows:

```
Command: _Helix
Number of turns = 3 Twist = CCW
Specify center point of base: enter 160,160
  right-click
Specify base radius or [Diameter] <1>: pick
  160,200
```

Fig. 14.37 First example –
Helix – the resulting helix

```
Specify top radius or [Diameter] <1>: enter 40
   right-click
Specify helix height or [Axis endpoint/Turns/turn
   Height/tWist] <1>: enter t (Turns)
   right-click
Enter number of turns <3>: enter 10 right-click
Specify helix height or [Axis endpoint/Turns/turn
   Height/tWist] <1>: enter 100 right-click
Command:
```

4. Call the **Extrude** tool form the **3D Modeling** panel and extrude the outline along the path of the helix. The command line shows:

```
Command: _extrude
Current wire frame density: ISOLINES=4
Select objects to extrude: pick the outline 1
   found
Select objects to extrude: right-click
Specify height of extrusion or [Direction/Path/
   Taper angle]: enter p (path) right-click
Select extrusion path or [Taper angle]: pick the
   helix
Command:
```

The result is shown in Fig. 14.37.

5. Add three cylinders, one to fit inside the helix, the second to form the shank of the screw, the third for the head of the screw. Subtract a box from the head for the screw slot. Then union the four parts of the screw.

6. Shade the screw using the **Conceptual** visual style.
   The result is shown in Fig. 14.38.

Fig. 14.38 First example –
Helix

Using Dynamic Input

As with all other tools (commands) in AutoCAD 2009 a helix can be formed working with the **Dynamic Input** (DYN) system. Figure 14.39 shows the stages (1 to 5) in the construction of the helix in the second example.

Set **DYN** on with a *click* on its button in the status bar.

1. *Click* the **Helix** tool icon in the **Home/3D Modeling** panel. The first of the **DYN** prompts appears. *Enter* **160,160** at the command line or drag the cursor until **160,160** appears in the **DYN** tip and *right-click*.
2. Move the cursor until the dimension **40** shows and *right-click*.
3. Press the *down arrow* key of the keyboard and *click* **Turns** in the menu which appears
4. *Enter* **6** and *right-click*.
5. Press the *down arrow* key of the keyboard and *enter* **100** in the menu as shown.

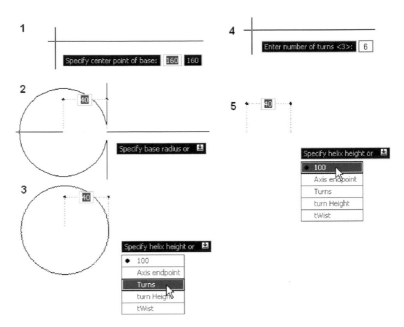

Fig. 14.39 Constructing the helix for the second example with the aid of **DYN**

3D Surfaces

As mentioned on p. 228, surfaces can be formed using the **Extrude** tool on lines and polylines. Two examples are given below in Figs 14.41 and 14.43.

First example – **3D Surface** (Fig. 14.41)

1. In the **ViewCube/Top** view, on a layer colour **Magenta**, construct the polyline shown in Fig. 14.40.
2. In the **ViewCube/Isometric** view, call the **Extrude** tool from the **Home/3D Modeling** control and extrude the polyline to a height of **80**. The result is shown in Fig. 14.41.

Second example – **3D Surface** (Fig. 14.43)

1. In the **Top** view on a layer colour **red** construct the circle shown in Fig. 14.42. Using the **Break** tool break the circle as shown.
2. In the **ViewCube/Isometric** view, call the **Extrude** tool and extrude the part circle to a height of **80**. Shade in the **Realistic** visual style.

The result is shown in Fig. 14.43.

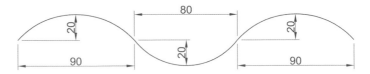

Fig. 14.40 First example – **3D Surface** – polyline to be extruded

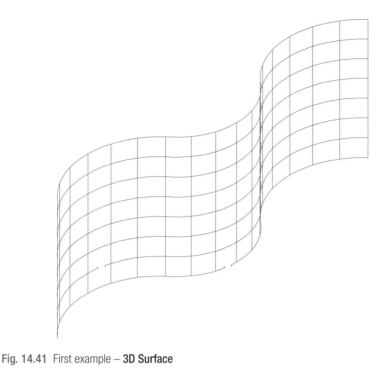

Fig. 14.41 First example – **3D Surface**

Fig. 14.42 Second example – **3D Surface** – the part circle to be extruded

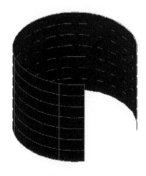

Fig. 14.43 Second example – **3D Surface**

REVISION NOTES

1. 3D models can be saved as blocks in a similar manner to the method of saving 2D drawings as blocks.
2. Libraries can be made up from 3D model drawings.
3. 3D models saved as blocks can be inserted into other drawings via the **DesignCenter**.
4. Arrays of 3D model drawings can be constructed in 3D space using the **3D Array** tool.
5. 3D models can be mirrored in 3D space using the **3D Mirror** tool.
6. 3D models can be rotated in 3D space using the **3D Rotate** tool.
7. 3D models can be cut into parts with the **Slice** tool.
8. Sectional views can be obtained from the 3D model using the **Section** tool.
9. Helices can be constructed using the **Helix** tool. The helices so formed can be used as paths for extruding outlines.
10. Both the **ViewCube** and **Viewpoint Presets** can be used for placing 3D models in different viewing positions in 3D space.
11. The **Dynamic Input** (**DYN**) method of construction can be used equally as well when constructing 3D model drawings as when constructing 2D drawings.
12. 3D surfaces can be formed from polylines or lines with **Extrude**.

CHAPTER 14

Exercises

Methods of constructing answers to the following exercises can be found in the free website:

http://books.elsevier.com/companions/9780750689830

1.   Figure 14.44 shows a shaded view of the 3D model for this exercise. Figure 14.45 shows a three-view projection of the model. Working to the details given in Fig. 14.45, construct the 3D model.

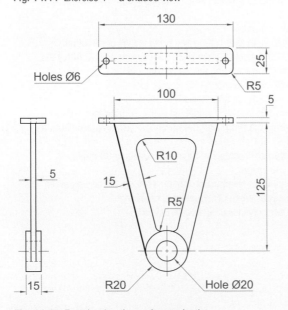

Fig. 14.44 Exercise 1 – a shaded view

Fig. 14.45 Exercise 1 – three-view projection

2.   Construct a 3D model drawing of the separating link shown in the two-view projection (Fig. 14.46). With the **Slice** tool, slice the model into two parts and remove the

rear part. Place the front half in an isometric view using the **ViewCube.** Shade the resulting model.

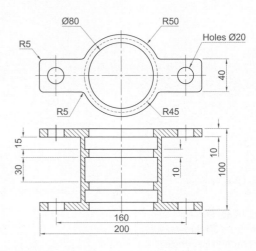

Fig. 14.46 Exercise 2

3.   Working to the dimensions given in the two orthographic projections (Fig. 14.48), and working on two layers of different colours, construct an assembled 3D model of the one part inside the other. With the **Slice** tool, slice the resulting 3D model into two equal parts, place in an isometric view and call the **Hide** tool as indicated in Fig. 14.47. Shade the resulting model in **Realistic** mode.

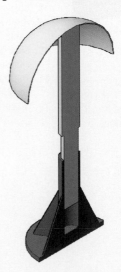

Fig. 14.47 Exercise 3

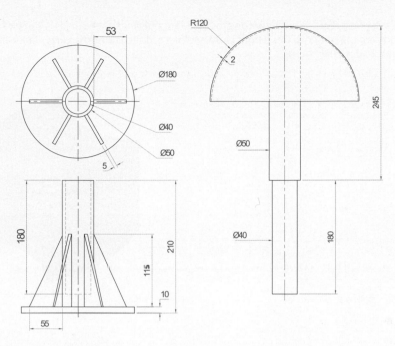

Fig. 14.48 Exercise 3 – orthographic projections

4.   Construct a solid of revolution of the jug shown in the orthographic projection (Fig. 14.49). Construct a
      handle from an extrusion of a circle along a semicircular path. Union the two parts. Place the 3D model in
      a suitable isometric view and render.

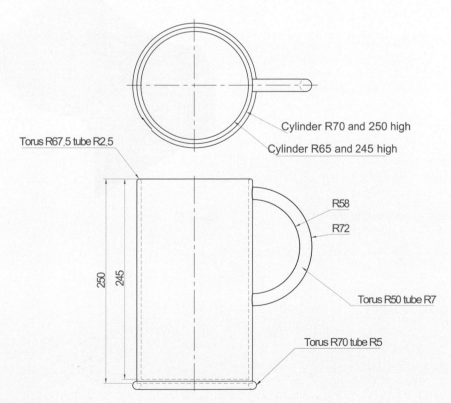

Fig. 14.49 Exercise 4

5.   In the **Top** view on a layer colour **blue** construct the four polylines shown in Fig. 14.50. Call the **Extrude** tool and extrude the polylines to a height of **80** and place in the **Isometric** view. Then call **Visual Styles/ Conceptual** shading (Fig. 14.51).

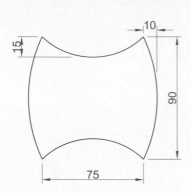

Fig. 14.50 Exercise 5 – outline to be extruded

Fig. 14.51 Exercise 5

6.   In the **Right** view construct the lines and arc shown in Fig. 14.52 on a layer colour **green**. Extrude the lines and arc to a height of **180**, place in the **Isometric** view and in the shade style **Visual Styles/ Realistic** (Fig. 14.53).

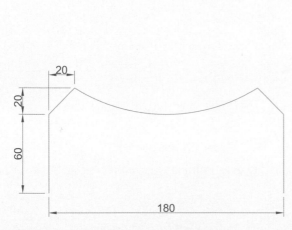

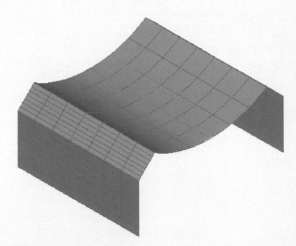

Fig. 14.52 Exercise 6 – outline to be extruded

Fig. 14.53 Exercise 6

Chapter 15

Rendering

AIMS OF THIS CHAPTER

The aims of this chapter are:

1. to construct a template for 3D modelling to be used as the drawing window for further work in 3D in this book;
2. to introduce the use of the **Render** tools in producing photographic-like images of 3D solid models;
3. to show how to illuminate a 3D solid model to obtain good lighting effects when rendering;
4. to give examples of the rendering of 3D solid models;
5. to introduce the idea of adding materials to 3D solid models in order to obtain a realistic appearance to a rendering;
6. to demonstrate the use of the forms of shading available using **Visual Styles** shading;
7. to demonstrate methods of printing rendered 3D solid models.

Setting up a new 3D template

In this chapter we will be constructing all 3D model drawings in the **acadiso3D.dwt** template.

1. *Click* the **Workspace Switching** button and *click* **3D Modeling** from the menu which appears (Fig. 15.1).
2. The AutoCAD window (Fig. 15.2) appears.

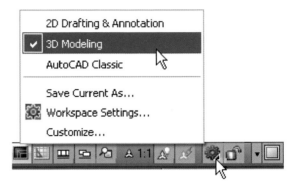

Fig. 15.1 *Click* **3D Modeling** in the **Workspace Settings menu**

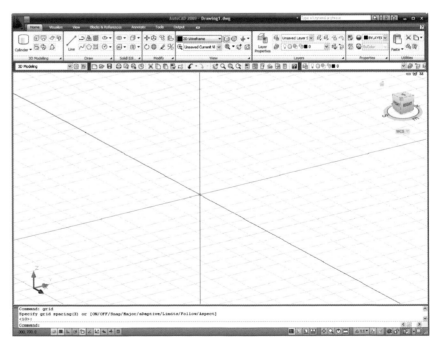

Fig. 15.2 The **3D Modeling** workspace

3. In the Command palette *enter*:

```
Command: enter perspective <1>: enter 0
Command:
```

4. Open the **Options** dialog with a *right-click* in the **Command palette**, followed by a *click* on **Options…** in the menu which appears.

5. *Click* the **Colors** tab. The **Drawing Window Colors** dialog appears. Set **Uniform Background** colour to **White** (Fig. 15.3).

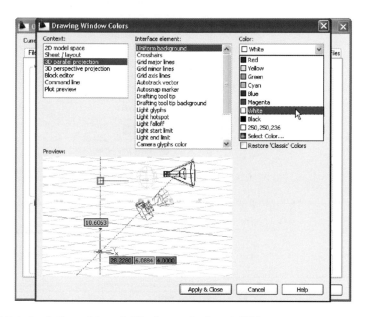

Fig. 15.3 In the **Options** dialog set all **background** colours to **White**

Note

The reader may prefer to retain the original background colours from the **Options** dialog. The colour **White** for backgrounds is used in this book for the clarity of illustrations.

6. Set **Units** to a **Precision** of 0, **Snap** to 5 and **Grid** to 10. Set **Limits** to 420,29. **Zoom** to **All**.
7. In the **Options** dialog *click* the **Files** label and *click* **Default Template File Name** for QNEW (Fig. 15.4) followed by a *click* on the **Browse** button which brings up the **Select Template** dialog, from which the acadiso3d.dwt can be selected. Now when AutoCAD 2009 is opened from the desktop, the **acadiso3D.dwt** template will open.
8. Set up five layers of different colours. In **my template** these have been named after the colours (Fig. 15.5).
9. Save the template to the name **acadiso3D** and then *enter* a suitable description in the **Template Definition** dialog.

Note

Figure 15.2 shows the screen is in **Parallel** projection. Some operators may prefer the screen to be set to **Perspective** projection.

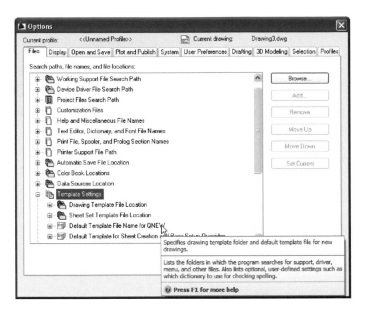

Fig. 15.4 Setting the default template setting in the **Options** dialog

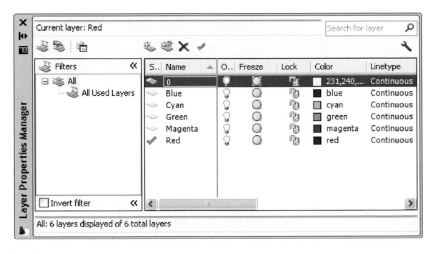

Fig. 15.5 Set up five new layers

Palettes

Click the **Tool Palettes** icon in the **View/Palettes** panel (Fig. 15.6). The **Tool Palettes – All Palettes** palette appears *docked* at the right-hand edge

Fig. 15.6 *Click* the **Tool Palettes** icon

of the AutoCAD window. *Drag* the palette away from its *docked* position. *Right-click* in the title bar of the palette and a *right-click* menu appears (Fig. 15.7). In Fig. 15.7 the **Draw** palette is showing.

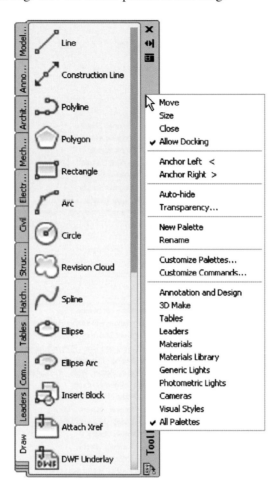

Fig. 15.7 The **Tool Palettes – All Palettes** palette with its *right-click* menu

To choose the tool palette required for the work in hand, either *click* on the name of the required palette in the tabs at the right-hand side of **All Palettes** palette or, if the required name is not showing, select the palette name from the *right-click* menu appearing with a *right-click* in the bottom left-hand corner of the **All Palettes** palette (Fig. 15.8).

The **All Palettes** palette can be *docked* against the side of the AutoCAD window if needed.

Applying materials to a model

Materials can be applied to a 3D model directly from icons in a selected palette. Three examples follow – applying a masonry material, applying a wood material and applying a metal material.

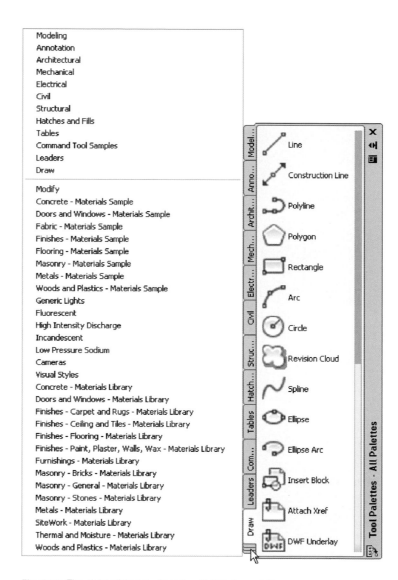

Fig. 15.8 The *right-click* menu from the **All Palettes** palette

Examples of applying materials

In the three examples which follow, lighting effects are obtained by turning **Sun Status** on, by *clicking* the **Sun Status** icon in the **Visualize/Sun** panel (Fig. 15.9). The **Lighting – Viewport Lighting Mode** dialog appears (Fig. 15.10). *Click* **Turn off the default lighting (recommended)**.

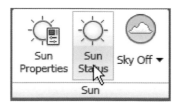

Fig. 15.9 The **Sun Status** icon in the **Visualize/Sun** panel

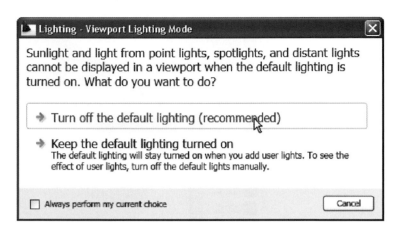

Fig. 15.10   The **Lighting – Viewport Lighting Mode** dialog

When the material has been applied, *click* the **Render** icon in the **Output/ Render** panel (Fig. 15.11) and the model will render.

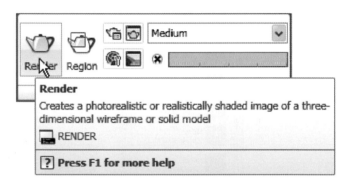

Fig. 15.11   The **Render** icon from the **Output/Render** panel

First example – applying a masonry material (Fig. 15.12)

Construct the necessary 3D model. In the **All Palettes** palette *click* the tab labelled **Masonry – Materials Sample**. *Right-click* on the icon representing the material to be applied to the model and from the menu which appears, *click* **Apply Materials To Objects**. A small icon similar to a paint brush appears with a small square. Move the square until it is on part of the model to which the material is to be applied and *left-click*, followed by a *right-click*. Then render the model. Figure 15.12 shows the resulting rendering.

Second example – applying a metal material (Fig. 15.13)

Construct the necessary 3D model. From the **All Palettes** palette *click* the tab labelled **Metals – Materials Sample**. *Right-click* on the icon representing the material to be applied to the model and from the menu which appears, *click* **Apply Materials To Objects**. *Left-click* the model to be rendered, followed by a *right-click* and render the model. The rendering is shown in Fig. 15.13.

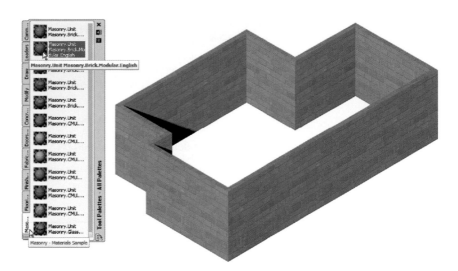

Fig. 15.12 First example – applying a masonry material

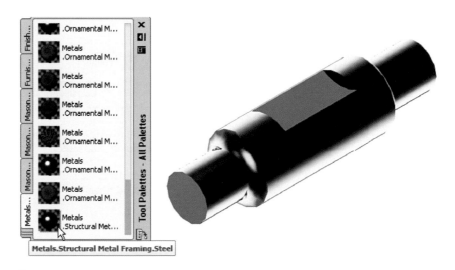

Fig. 15.13 Second example – applying a metal material

Third example – applying a wood material (Fig. 15.14)

Construct the necessary 3D model. In the **All Palettes** palette *click* the tab labelled **Wood and Plastics – Materials Sample**. *Right-click* on the icon representing the material to be applied to the model and from the menu which appears, *click* **Apply Materials To Objects**. *Left-click* the model to be rendered, followed by a *right-click* and render the model. The rendering is shown in Fig. 15.14.

Modifying an applied material

If the result of applying a material direct to a model from the selected materials palette is not satisfactory, modifications to the applied material can be made. In the case of the second example, in the **Visualize/ Materials** panel *right-click* the **Materials** icon (Fig. 15.15) and the **Materials** palette appears, showing the **Available Materials in Drawing** palette (Fig. 15.16). Features such as colour of the applied material, tiling, and different texture maps of the material (or materials) applied to a model can be amended as wished from this palette.

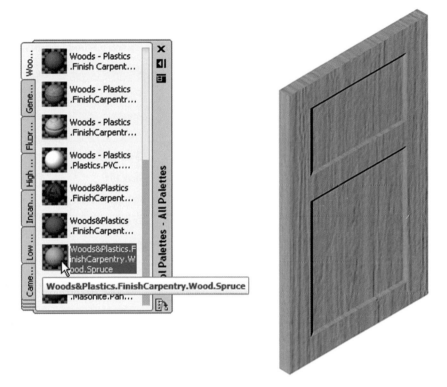

Fig. 15.14 Third example – applying a wood material

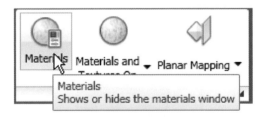

Fig. 15.15 The **Materials** icon in the **Visualize/Materials** panel

Fourth example – **Available Materials in Drawing** – Fig. 15.17

As an example, Fig. 15.17 shows the nine materials applied to various parts of a 3D model of a hut in a set of fields surrounded by fences. The **Available Materials in Drawing** are shown, together with the **Material Editor** for the top-left-hand material (outlined in yellow). A *click* on a material in the **Available Materials in Drawing** brings up details of that selected material.

Fig. 15.16 The **Materials** palette showing the available materials in drawing

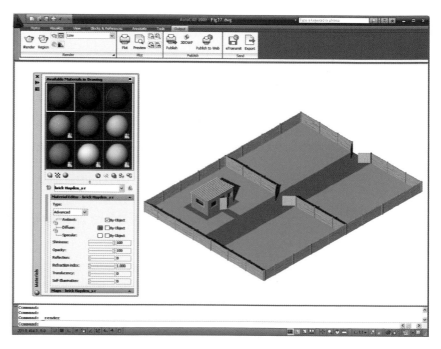

Fig. 15.17 Fourth: example of materials applied to parts of a 3D model

The **Render** tools and dialogs

The tool icons and menus in the **Output/Render** panel are shown in Fig. 15.18.

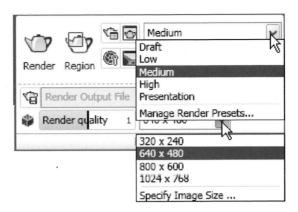

Fig. 15.18 The tools and menus in the **Output/Render** panel

The **Lights** tools

The different forms of lighting from light palettes are shown in Fig. 15.19. There are many forms of lighting available when using AutoCAD 2009; the most frequently used include the following:

- **Ambient** lighting is taken as the general overall light that is all around and surrounding any object. Usually left at 30%.

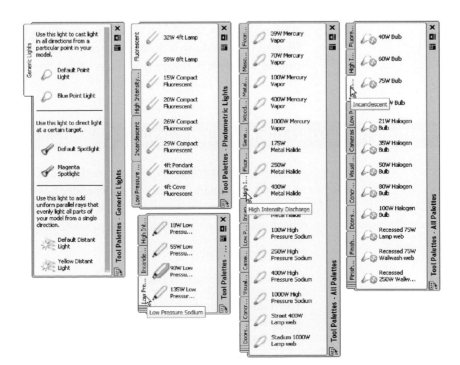

Fig. 15.19 The **Lights** tools palettes

CHAPTER 15

- **Point** lights shed light in all directions from the position in which the light is placed.
- **Distant** lights send parallel rays of light from their position in the direction chosen by the operator.
- **Spotlights** illuminate as if from a spotlight. The light is in a direction set by the operator and is in the form of a cone, with a 'hotspot' cone giving a brighter spot on the model being lit.
- **Photometric lighting** is lighting in which lights of a selected wattage can be placed in a lighting scene. The set variable **LIGHTINGUNITS** must be set to **1** or **2** for photometric lights to function.
- **Viewport lighting mode** in **Default lighting** or **User light/sunlight**.
- **Sun** light, which can be edited.
- **Sky background and illumination**.

Other forms of lighting are shown in Fig. 15.19. In this book, examples of lighting methods shown in examples will only be concerned with the use of **Point** and **Direct** lights, together with **Default lighting**.

Placing lights to illuminate a 3D model

Any number of the three types of lights – **Point**, **Distant** and **Spotlight** – can be positioned in 3D space as wished by the operator.

In general, good lighting effects can be obtained by placing a **Point** light high above the object(s) being illuminated, with a **Distant** light placed pointing towards the object at a distance from the front and above the general height of the object(s) and with a second **Distant** light pointing towards the object(s) from one side and not as high as the first **Distant** light. If desired **Spotlights** can be used either on their own or in conjunction with the other two forms of lighting.

Setting rendering background colour

The default background colour for rendering in the **acadiso 3D** template is a grey by default. In this book, all renderings are shown on a white background in the viewport in which the 3D model drawing was constructed. To set the background to white for renderings:

1. At the command line:

Command: *enter* view *right-click*

The **View Manager** dialog appears (Fig. 15.20). *Click* **Model View** in its **Views** list, followed by a *click* on the **New...** button.

2. The **New View/Shot Properties** dialog (Fig. 15.21) appears. *Enter* **current** (or similar) in the **View name** field. In the **Background** pop-up list *click* **Solid**. The **Background** dialog appears (Fig. 15.22).

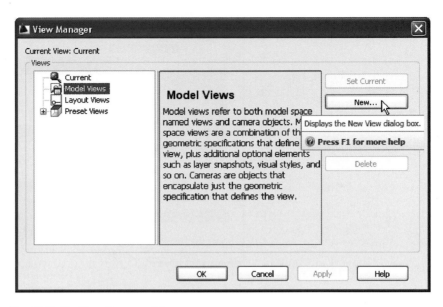

Fig. 15.20 The **View Manager** dialog

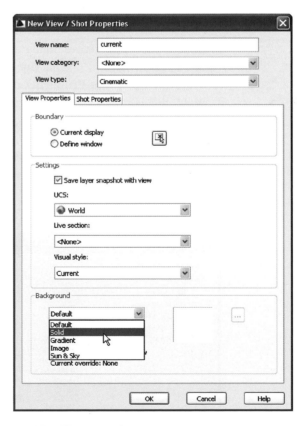

Fig. 15.21 The **New View/Shot Properties** dialog

3. In the **Background** dialog *click* in the **Color** field. The **Select Color** dialog appears (Fig. 15.22).
4. In the **Select Color** dialog *drag* the slider as far upwards as possible to change the colour to white (**255,255,255**). Then *click* the dialog's **OK**

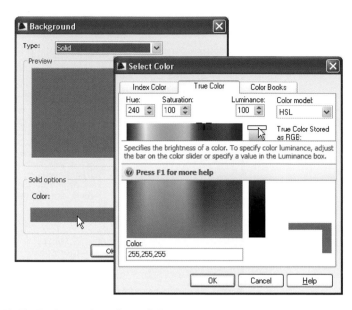

Fig. 15.22 The **Background** and **Select Color** dialogs

button. The **Background** dialog reappears, showing white in the **Color** and **Preview** fields. *Click* the **Background** dialog's **OK** button.

5. The **New View/Shot Properties** dialog reappears, showing **current** highlighted in the **Views** list. *Click* the dialog's **OK** button.
6. The **View Manager** dialog reappears. *Click* the **Set Current** button, followed by a *click* on the dialog's **OK** button (Fig. 15.23).
7. *Click* the **Advanced Render Settings...** icon (Fig. 15.24). The **Advanced Render Settings** palette appears.

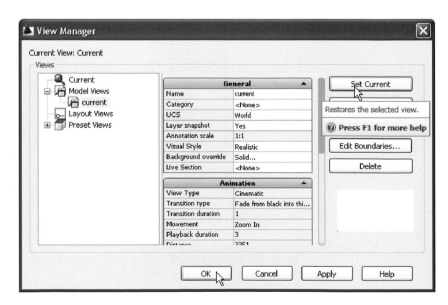

Fig. 15.23 The **View Manager** dialog reappears

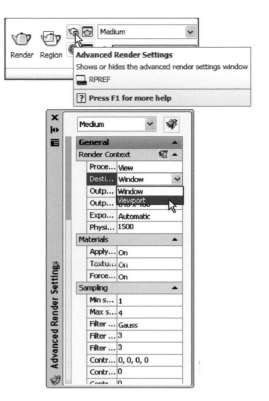

Fig. 15.24 The **Advanced Render Settings** icon and the **Advanced Render Settings** palette

8. In the palette *click* in the **Render Context** field, *click* the arrow to the right of **Window** and in the pop-up menu which appears *click* **Viewport** as the rendering destination (Fig. 15.24).
9. Close the palette and save the screen with the new settings as the template **acadiso3D.dwt**. This will ensure that renderings are made in the viewport in which the 3D model was constructed to be the same viewport in which renderings are made – on a white background.

First example – Rendering a 3D model (Fig. 15.37)

1. Construct a 3D model of the wing nut shown in the two-view projection of Fig. 15.25.
2. Place the 3D model in the **ViewCube/Top** view (Fig. 15.26), **Zoom** to **1** and with the **Move** tool, move the model to the upper part of the AutoCAD drawing area.
3. *Click* the **Point Light** tool icon in the **Visualize/Lights** panel (Fig. 15.27). The warning window (Fig. 15.28) appears. *Click* **Turn off Default Lighting** in the window.
4. A **New PointLight** icon appears and the command line shows:

```
Command: _pointlight
Specify source location <0,0,0>: enter .xy
  right-click
```

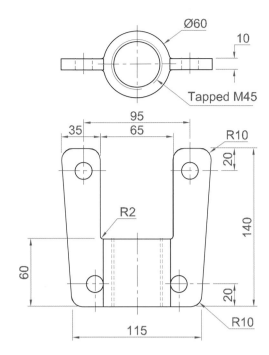

Fig. 15.25 First example – Rendering a 3D model – two-view projection

Fig. 15.26 Place the model in the **ViewCube/Top** view

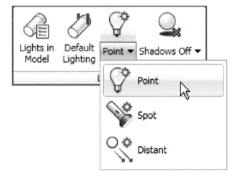

Fig. 15.27 The **Point Light** icon in the **Visualize/Lights** panel

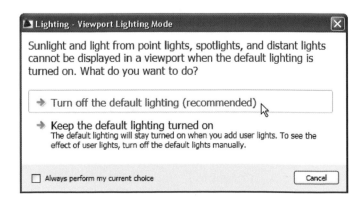

Fig. 15.28 The Lighting – **Viewport Lighting Mode** warning window

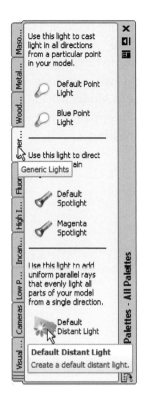

```
of click at center of model (need Z): enter 500
    right-click
Enter an option to change [Name/Intensity/Status/
    shadoW/Attenuation/Color/eXit]
<eXit>: enter n (Name) right-click
Enter light name <Pointlight1>: right-click
Enter an option to change [Name/Intensity/
    Status/shadoW/Attenuation/Color/eXit] <eXit>:
    right-click
Command:
```

5. There are several methods by which **Distant** lights can be called: by selecting **Default Distant Light** from the **Generic Lights** palette (Fig. 15.29); with a *click* the on the **Distant** icon in the **Vizualise/Lights** panel; or by *entering* **distantlight** at the command line.

No matter which method is adopted, the **Lighting – Photometric Distant Lights** dialog (Fig. 15.30) appears. *Click* **Allow Distant lights** and the command line shows:

```
Command: _distantlight
Specify light direction FROM <0,0,0> or [Vector]:
    enter .xy right-click
of click to the left and below the 3D model (need
    Z) enter 400 right-click
Specify light direction TO <1,1,1>: enter .xy
    right-click
of click at center of mode (need Z) enter 70
    right-click
Command:
```

Fig. 15.29 The **Generic Lights** palette showing the **Default Distant Light** selected

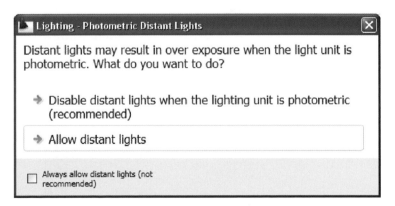

Fig. 15.30 The **Lighting – Photometric Distant Lights** dialog

6. Place another **Distant Light** (**Distantlight2**) in the same position **TO** and **FROM** the front and below the model at **Z** of **300**.

7. When the model has been rendered if a light requires to be changed in intensity, shadow, position or colour, *click* the **Light List** icon in the **Visualize/Lights** panel (Fig. 15.31) and the **Lights in Model** palette appears. *Double-click* a light name and the **Properties** palette for the light appears in which modifications can be made (Fig. 15.32). Amendments can be made as thought necessary.

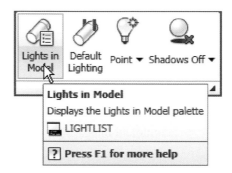

Fig. 15.31 Selecting the **Lights** icon from the **Visualize/Lights** panel

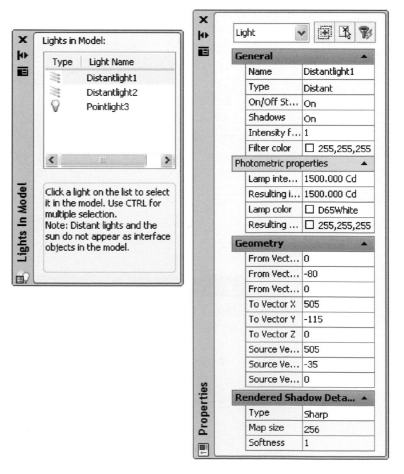

Fig. 15.32 The **Lights in Model** and **Properties** palettes

Note

In this example the **Intensity factor** has been set at **1** for lights. This is possible because the lights are close to the model. In larger size models the **Intensity factor** may have to be set to a higher figure.

Adding a material to the model

1. In the **Metals – Materials Sample** palette select **Metals, Structural and Metal Framing Steel** (Fig. 15.33) to apply to the 3D model.
2. *Click* The **Materials** icon. The **Materials** palette appears showing an icon of the applied material (Fig. 15.34). In the palettes *click* the

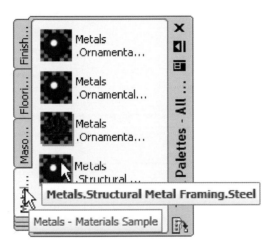

Fig. 15.33　The **Metals, Structural Framing, Steel** icon

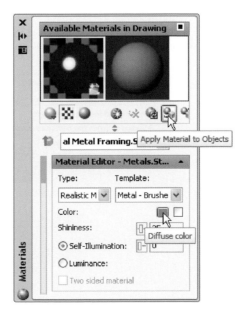

Fig. 15.34　The **Materials** pallete showing the applied material

Diffuse color button and from the **Select Color** dialog which appears select a suitable colour for the model.

3. *Click* **Apply Material to Objects** icon (Fig. 15.34).
4. *Click* any part of the 3D model to apply the material to the model.
5. *Right-click* in the **Type** field of the **Material Editor** section of the palette and select **Realistic** in the *right-click* menu (Fig. 15.35).

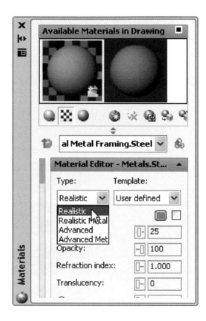

Fig. 15.35 Select **Realistic** in the **Type** drop-down menu

6. *Click* the **Advanced Render Settings** tool in the **Output/Render** toolbar and in the palette which appears select **Presentation** from the list at the top of the palette (Fig. 15.36).
7. Render the 3D model again and if now satisfied save to a suitable file name.

Figure 15.37 shows four renderings in the four **Type** settings.

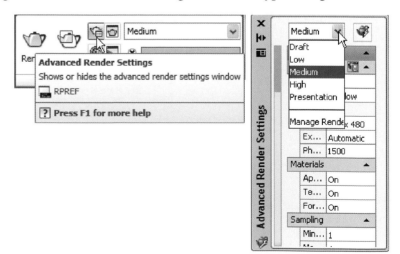

Fig. 15.36 Setting the form of rendering to **Presentation**

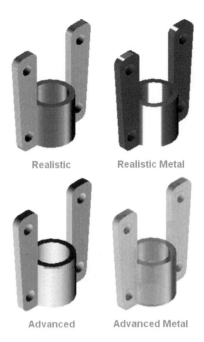

Realistic   Realistic Metal

Advanced   Advanced Metal

Fig. 15.37 First example – Rendering a 3D model

> **Note**
>
> The limited descriptions of rendering given in these pages does not
> show the full value of different types of lights, materials and rendering
> methods. The reader is advised to experiment with the facilities
> available for rendering.

Second example – Rendering a 3D model (Fig. 15.39)

1. Construct 3D models of the two parts of the stand and support given in the
   projections shown in Fig. 15.38 with the two parts assembled together.

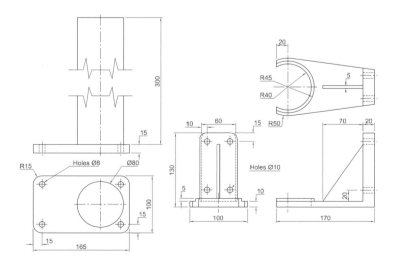

Fig. 15.38 Second example – Rendering – orthographic projection

Fig. 15.39 Second example – Rendering

2. Place the scene in the **ViewCube/Top** view, **Zoom** to **1** and add lighting.
3. Add different materials to the parts of the assembly and render the result.

Figure 15.39 shows the resulting rendering.

Third example – Rendering a 3D model (Fig. 15.40)

Figure 15.40 shows an exploded, rendered 3D model of a pumping device from a machine and Fig. 15.41 a third-angle orthographic projection of the device from a machine.

Fig. 15.40 Third example – Rendering

The **3dorbit** tool

At the command line *enter* **3dorbit**. The command line shows:

```
Command: 3dorbit
```

Press ESC or ENTER to exit, or *right-click* to display the shortcut menu.

Right-click anywhere on screen and the **3dorbit** *right-click* menu appears (Fig. 15.42). *Click* **Free Orbit**. A circle and movement icon appears on screen. The position and angle of the model can be adjusted by either *clicking* in one of the four outer small circles or by *clicking* outside the main circle and moving the mouse.

Example – **3dorbit** (Fig. 15.42)

This is another tool for the manipulation of 3D models into different positions within 3D space.

1. Open the file of the second example of rendering (Fig. 15.38).
2. Shade the model using **Visual Styles/Realistic**.
3. *Enter* **3dorbit** at the command line.
4. With the cursor outside the circle move the mouse. The 3D model rotates within the circle.

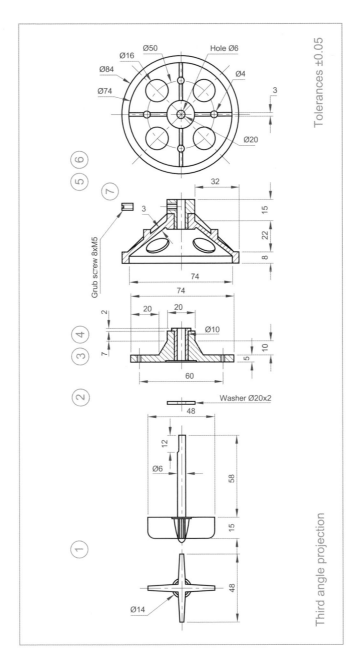

Fig. 15.41 Third example – Rendering a 3D model – exploded orthographic views

5. With the cursor inside the circle move the mouse. The 3D model rotates around the screen.
6. With the cursor inside any one of the small quadrant circles the 3D model can be moved vertically or horizontally as the mouse is moved (Fig. 15.41).
7. Fit the 3D model into a **Four: Equal** viewports setting. Note the **Realistic** visual style still shows in each of the four viewports and that the **3dorbit** tool can still be used as shown in the bottom-right-hand viewport (Fig. 15.43).

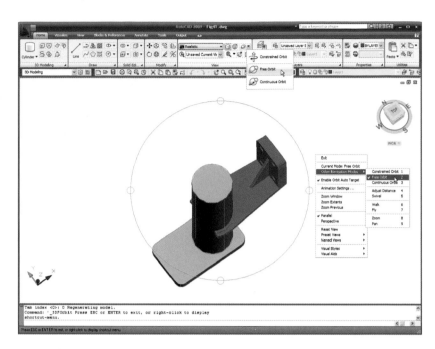

Fig. 15.42 The *right-click* menus of the **3dorbit** tool

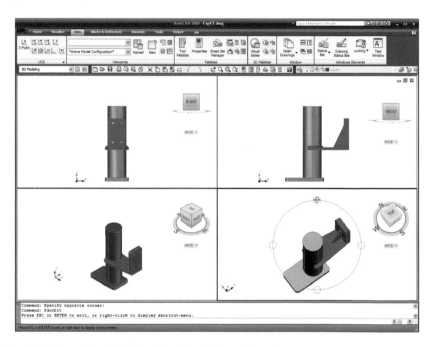

Fig. 15.43 Example – **3dorbit** in a **Four: Equal** viewport layout

Producing hard copy

Printing or plotting a drawing on screen using AutoCAD 2009 can be carried out either from **Model Space** or from **Paper Space**. In versions of AutoCAD before AutoCAD 2004, it was necessary to print or plot from **Pspace**.

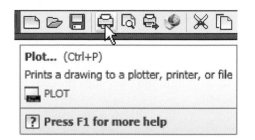

Fig. 15.44 Calling the **Plot** tool from the **Standard** toolbar

First example – printing a single copy (Fig. 15.45)

> **Note**
>
> The drawing being printed in this example is in a **Visual Styles/ Conceptual** shading mode.
>
> 1. With a drawing to be printed or plotted on screen *click* the **Plot** tool icon in the **Standard** toolbar (Fig. 15.44).
>
> 2. The **Plot** dialog appears. Set the **Printer/Plotter** to a printer or plotter currently attached to the computer and the **Paper Size** to a paper size to which the printer/plotter is set.
>
> 3. *Click* the **Preview** button of the dialog and if the preview is OK, *right-click* and in the *right-click* menu which appears, *click* **Plot**. The drawing plots producing the necessary hard copy (Fig. 15.45).

Fig. 15.45 First example – printing a single copy

Second example – multiple view copy (Fig. 15.46)

A 3D model to be printed is a **Realistic** view of a 3D model which has been constructed on three layers – **Red**, **Blue** and **Green** in colour. To print a multiple view copy proceed as follows:

1. Place the drawing in a **Four: Equal** viewport setting.
2. Make a new layer **vports** of colour cyan and make it the current layer.
3. *Click* the **Layout** button in the status bar. The drawing appears in **Pspace**. A view of the 3D model appears within a cyan coloured viewport (Fig. 15.46).
4. *Click* the **Plot** tool icon in the **Output/Plot** toolbar. Make sure the correct **Printer/Plotter** and **Paper Size** settings are selected and *click* the **Preview** button of the dialog.
5. A preview of the 3D model appears.
6. If the preview is satisfactory (Fig. 15.46), *right-click* and from the right-click menu *click* **Plot**. The drawing plots to produce the required four-viewport hard copy.

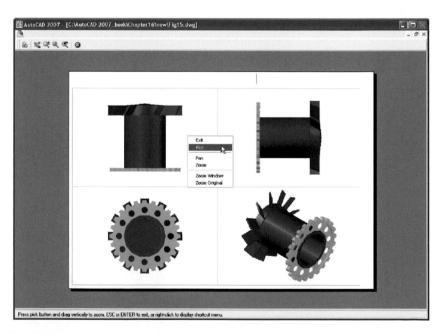

Fig. 15.46 Second example – multiple view copy

Other forms of hard copy

When working in AutoCAD 2009, several different forms of hard copy in **Home/View** visual styles are possible. As an example a single view plot review of the same 3D model is shown in the **Hidden** shading form (Fig.15.47).

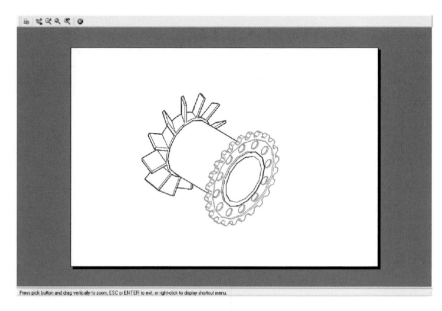

Fig. 15.47 An example of a **Hidden Style** plot **Preview**

Saving and opening 3D model drawings

3D model drawings are saved and/or opened in the same way as are 2D
drawings. To save a drawing *click* **Save As...** in the **File** drop-down menu
and save the drawing in the **Save Drawing As** dialog and *enter* a file name
in the **File Name** field of the dialog before *clicking* the **Save** button. To
open a drawing which has been saved *click* **Open...** in the **File** drop-down
menu, and in the **Select File** dialog which appears select a file name from
the file list.

There are differences between saving a 2D and a 3D drawing, in that when
3D model drawing is shaded by using a visual style from the **Home/View**
panel, the shading is saved with the drawing.

Exercises

Methods of constructing answers to the following exercises can be found in the free website:

http://books.elsevier.com/companions/9780750689830

1.   A rendering of an assembled lathe tool holder is shown in Fig. 15.48. The rendering includes different materials for each part of the assembly. Working to the dimensions given in the parts orthographic drawing of Fig. 15.49, construct a 3D model drawing of the assembled lathe tool holder on several layers of different colours, add lighting and materials and render the model in an isometric view. Shade with **3D Visual Styles/Hidden** and print or plot a **ViewCube/Isometric** view of the model drawing.

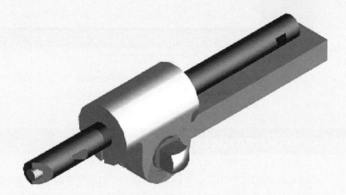

Fig. 15.48 Exercise 1

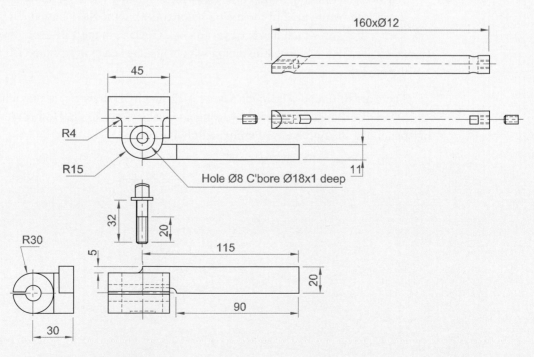

Fig. 15.49 Exercise 1 – parts drawings

2. Figure 15.50 is a rendering of a drip tray. Working to the sizes given in Fig. 15.51, construct a 3D model drawing of the tray. Add lighting and a suitable material, place the model in an isometric view and render.

3. A three-view drawing of a hanging spindle bearing in third angle orthographic projection is shown in Fig. 15.52. Working to the dimensions in the drawing construct a 3D model drawing of the bearing. Add lighting and a material and render the model.

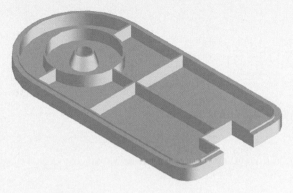

Fig. 15.50 Exercise 2

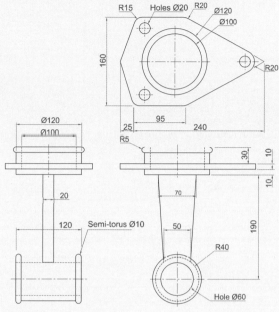

Fig. 15.52 Exercise 3

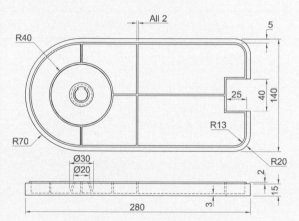

Fig. 15.51 Exercise 2 – two-view projection

Building drawings

AIMS OF THIS CHAPTER

AIMS OF THIS CHAPTER

The aims of this chapter are:

1. to show that AutoCAD 2009 is a suitable computer-aided design software package for the construction of building drawings;
2. to show that AutoCAD 2009 is a suitable CAD program for the construction of 3D models of buildings.

Building drawings

There are a number of different types of drawings related to the construction of any form of building. In this chapter a fairly typical example of a set of building drawings is shown. These are seven drawings related to the construction of an extension to an existing two-storey house (44 Ridgeway Road). These show:

1. a site plan of the original two-storey house, drawn to a scale of **1:200** (Fig. 16.1);
2. a site layout plan of the original house, drawn to a scale of **1:100** (Fig. 16.2);
3. floor layouts of the original house, drawn to a scale of **1:50** (Fig. 16.3);
4. views of all four sides of the original house drawn to a scale of **1:50** (Fig. 16.4);
5. floor layouts including the proposed extension, drawn to a scale of **1:50** (Fig. 16.5);
6. views of all four sides of the house including the proposed extension, drawn to a scale of **1:50** (Fig. 16.6);
7. a sectional view through the proposed extension, drawn to a scale of **1:50** (Fig. 16.7).

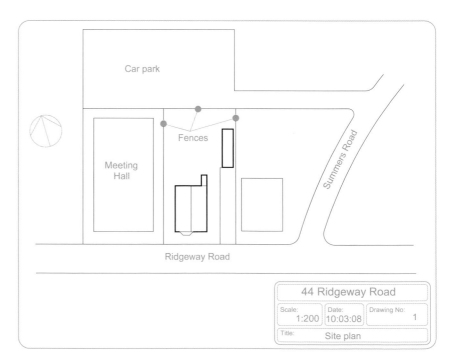

Fig. 16.1 A site plan

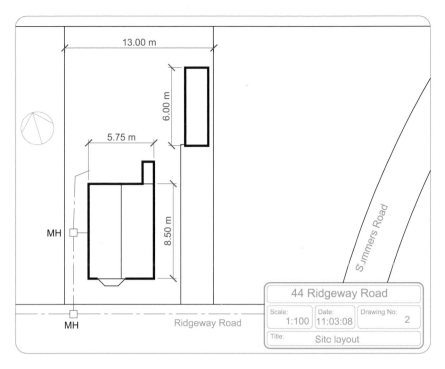

Fig. 16.2 A site layout plan

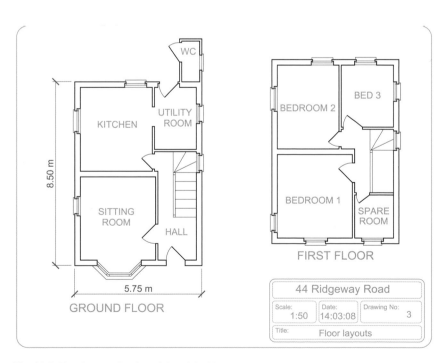

Fig. 16.3 Floor layouts drawing of the original house

Fig. 16.4 Views of the original house

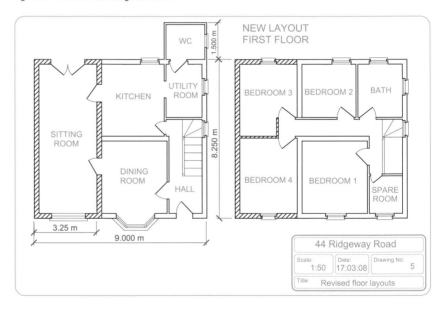

Fig. 16.5 Floor layouts drawing of the proposed extension

Note

1. Other types of drawing will be constructed such as drawings showing the details of parts such as doors, windows, and floor structures. These are often shown in sectional views.

2. Although the seven drawings related to the proposed extension of the house at 44 Ridgeway Road are shown here as having been constructed on either A3 or A4 layouts, it is common practice to include several types of building drawing on larger sheets such as A1 sheets of a size 820 mm by 594 mm.

Fig. 16.6 Views including the proposed extension

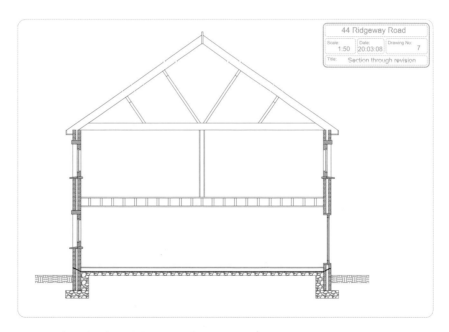

Fig. 16.7 A section through the proposed extension

Floor layouts

When constructing floor layout drawings it is advisable to build up
a library of block drawings of symbols representing features such as
doors and windows. These can then be inserted into layouts from the
DesignCenter. A suggested small library of such block symbols is shown
in Fig. 16.8.

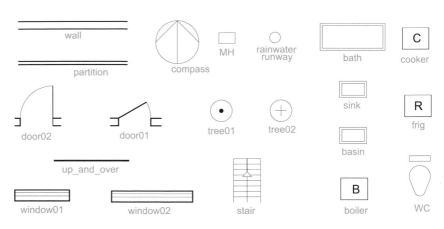

Fig. 16.8 A small library of building symbols

3D Models of buildings

Details of shapes and dimensions for the first two examples are taken from the drawings of the building and its extension at 44 Ridgeway Road given in the Figs. 16.2 to 16.6 on pages 305 to 307.

First example-44 Ridgeway Road-original building

Details of this first example are taken from Figures 16.2 to 16.4 on pages 305 and 306.

The following steps describe the construction of a 3D model of 44 Ridgeway Road prior to the extension being added.

1. In the **Layer Properties Manager** palette – **Doors** (colour **red**), **Roof** (colour green), **Walls** (colour blue), **Windows** (colour 8). Fig. 16.9.
2. Set the screen to the **ViewCube/Front** view (Fig. 16.10)

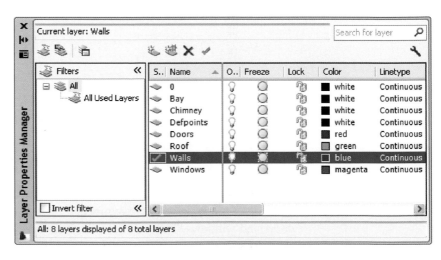

Fig. 16.9 First example – the layers on which the model is to be constructed

Fig. 16.10 Set screen to the
View Block/Front view

Fig. 16.11 First example – the walls

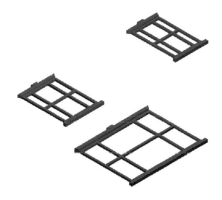

Fig. 16.12 First example – extrusions of the three sizes of window

3. Set the layer **Walls** current and, working to as scale of **1:50** construct outlines of the walls. Construct outlines of the bay, windows and doors inside the wall outlines.
4. **Extrude** the wall, bay, window and door outlines to a height of **1**.
5. **Subtract** the bay, window and door outlines from the wall outlines. The result is shown in Fig. 16.11.
6. Make the layer **Windows** current and construct outlines of three of the windows which are of different sizes. Extrude the copings and cills to a height of **1.5** and the other parts to a height of **1**. Form a union of the main outline, the coping and the cill. The window pane extrusions will have to be subtracted from the union. Fig. 16.12shows the 3D models of the three window in an **ViewCube/Isometric** view.
7. Move and copy the windows to their correct positions in the walls.
8. Make the layer **Doors** current and construct outlines of the doors and extrude to a height of **1**.
9. Make layer **Chimney** current and construct a 3D model of the chimney (Fig. 16.13).
10. Make the layer **Roofs** current and construct outlines of the roofs (main building and garage). See Fig. 16.14.
11. On the layer **Bay** construct the bay and its windows.

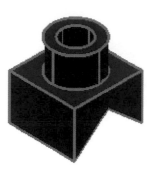

Fig. 16.13 First example –
Realistic view of a 3D model
of the chimney

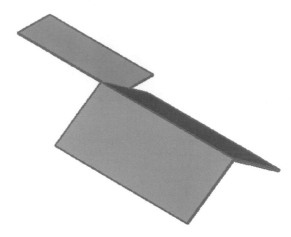

Fig. 16.14 First example – **Realistic** view of the roofs

Assembling the walls

1. Place the screen in the **ViewCube/Top** view (Fig. 16.15).
2. Make the layer **Walls** current and turn off all other layers other than **Windows**.
3. Placing a window around each wall in turn move and/or rotate and move the walls until they are in their correct position relative to each other.
4. Place in the **ViewCube/Isometric** view and using the move the walls into their correct positions relative to each other. Fig 16.16 shows the walls in position in a **ViewCube/Top** view.
5. Move the roof into position relative to the walls and move the chimney into position on the roof. Fig. 16.17shows the resulting 3D model in a **ViewCube/Isometric** view (Fig. 16.18).

Fig. 16.15 Set screen to **Top**
view

Fig.16.16 First example – the four walls in their correct positions relative to each other in a **View Block/Top** view

Fig. 16.17 First example – a **Realistic** view of the assembled walls, windows, bay, roof and chimney

Fig. 16.18 Set screen to a
ViewCube/Isometric view

Fig. 16.19 First example – **Realistic** view of the original house and garage

The garage

On layers **Walls** construct the walls and on layer **Windows** construct the
windows. Fig. 16.19 is a **Realistic** visual style view of the 3D model as
constructed so far,

Second example-extension to 44 Ridgeway Road

Working to a scale of **1:50** and taking dimensions from the drawing
Figures. 16.5 and 16.6 (page ??) and in a manner similar to the method of
constructing the 3D model of the original building, add the extension to
the original building. Fig. 16.20 shows a **Realistic** visual style view of the
resulting 3D model, In this 3D model floors have been added – a ground
and a second storey floor constructed on a new layer **Floors** of colour
yellow. Note the changes in the bay and front door.

Fig. 16.20 Second example – a **Realistic** view of the building with its extension

Third example-small building in fields

Working to a scale of **1:50** from the dimensions given in Fig. 16.21,
construct a 3D model of the hut following the steps given below.

The walls are painted concrete and the roof is corrugated iron.

In the **Layer Properties Manager** dialog make the new levels as follows:

Walls-colour **Blue**
Road-colour **Red**
Roof-colour **Red**
Windows-**Magenta**
Fence-colour **8**
Field-colour **Green**

Following the methods used in the construction of the house in the first
example, construct the walls, roof, windows and door of the small building
in one of the fields. Fig. 16.22 shows a **Realistic** visual style view of a 3D
model of the hut.

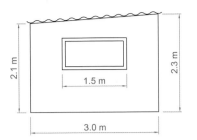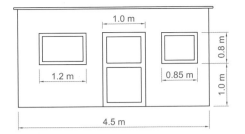

Fig. 16.21 Third example – front and end views of the hut

Fig. 16.22 Third example – **Realistic** view of a 3D model of the hut

Constructing the fence, fields and road

1. Place the screen in a **Four: Equal** viewports setting.
2. Make the **Garden** layer current and in the **Top** viewport, construct an outline of the boundaries to the fields and to the building. Extrude the outline to a height of **0.5**.
3. Make the **Road** layer current and in the **Top** viewport, construct an outline of the road and extrude the outline to a height of **0.5**.
4. In the **Front** view, construct a single plank and a post of a fence and copy them a sufficient number of times to surround the four fields leaving gaps for the gates. With the **Union** tool form a union of all the posts and planks. Fig. 16.23 shows a part of the resulting fence in a **Realistic** visual style view in the **Isometric** viewport. With the **Union** tool form a union of all the planks and posts in the entire fence.
5. While still in the layer **Fence**, construct gates to the fields.
6. Make the **Road** layer current and construct an outline of the road. Extrude to a height of **0.5**.

> **Note**
>
> When constructing each of these features it is advisable to turn off those layers on which other features have been constructed.

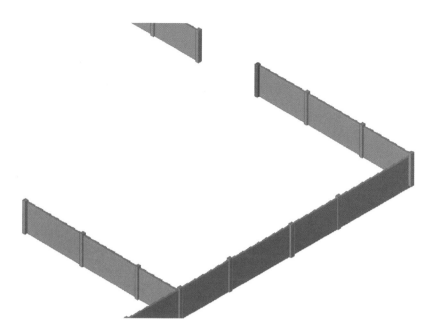

Fig. 16.23 Third example – part of the fence

Fig. 16.24 shows a **Conceptual** view of the hut in the fields with the road, fence and gates.

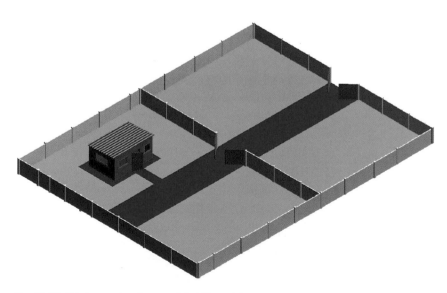

Fig. 16.24 Third example – the completed 3D model

Completing the second example

Working in a manner similar to the method used when constructing the roads, garden and fences for the third example, add the paths, garden area and fences and gates to the building 44 Ridgeway Road with its extension. Fig. 16.24 is a **Conceptual** visual style view of the resulting 3D model.

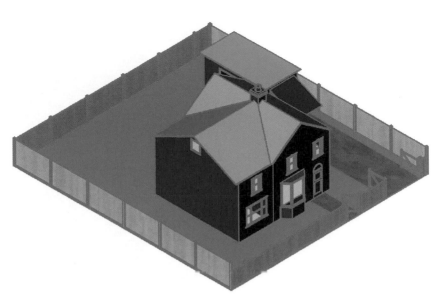

Fig. 16.25 Second example – the completed 3D model

Material attachments and rendering
Second example

The following materials were attached to the various parts of the 3D model
Fig. 16.25. To attach the materials, all layers except the layer on which
the objects to which the attachment of a particular material is being made
are tuned off, allowing the material in question to be attached only to the
elements to which each material is to be attached.

Default: colour **7.**
Doors: Wood Hickory
Fences: Wood-Spruce
Floors: Wood-Hickory
Garden: Green
Gates: Wood-White
Roofs: Brick-Herringbone
Windows: Wood-White

The 3D model was then rendered with **Output Size** set to **1024 × 768**
and **Render Preset** set to **Presentation**, with **Sun Status** turned on. The
resulting rendering is shown in Fig. 16.26.

Material attachments and rendering
Third example

Fig. 16.27 shows the third example after attaching materials and
rendering

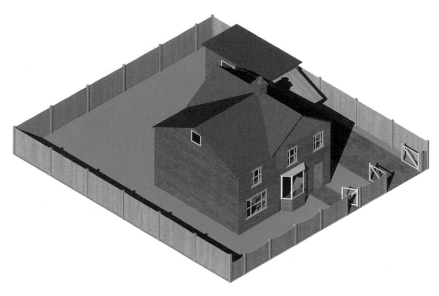

Fig. 16.26 Second example – a rendering after attaching materials

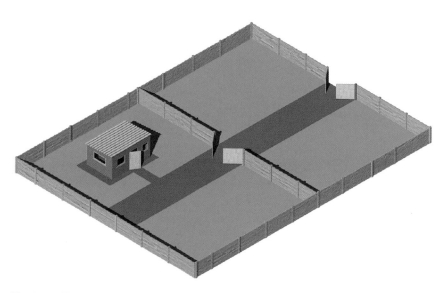

Fig. 16.27 Third example – 3D model after attaching materials and rendering

REVISION NOTES

1. There are a number of different types of building drawings-site plans, site layout plans, floor layouts, views, sectional views, detail drawings. AutoCAD 2009 is a suitable CAD programme to use when constructing building drawings.
2. AutoCAD 2009 is a suitable CAD programme for the construction of 3D models of buildings.

Exercises

Methods of constructing answers to the following exercises can be found in the free website: http://books. elsevier.com/companions/9780750689830

1.   Fig. 16.28 is a site plan drawn to a scale of 1:200 showing a bungalow to be built in the garden of an existing bungalow. Construct the library of symbols shown in Fig. 16.8 on page 308 and by inserting the symbols from the DesignCenter construct a scale 1:50 drawing of the floor layout plan of the proposed bungalow.

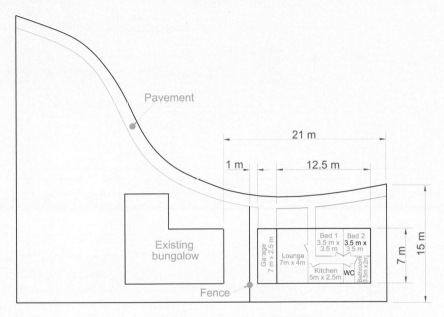

Fig. 16.28 Exercise 1

2.   Fig. 16.29 is a site plan of a two-storey house a building plot.Design and construct to a scale 1:50, a suggested pair of floor layouts for the two floors of the proposed house.

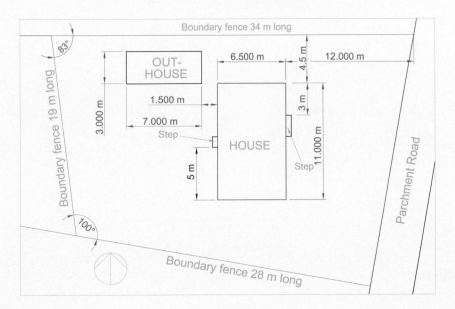

Fig. 16.29 Exercise 2

3.   Fig. 16.30 shows a scale 1:100 site plan for the proposed bungalow 4 Caretaker Road. Construct the floor layout for the proposed house shown in the drawing Fig. 16.28.

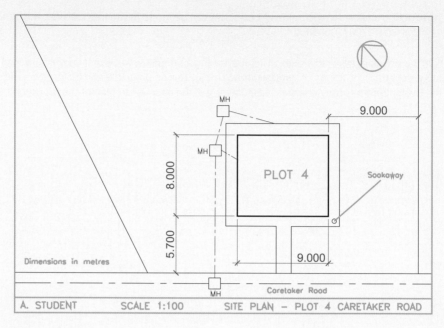

Fig. 16.30 Exercise 3 – site plan

4.   Fig. 16.31 shows views of a house in orthographic projection and Fig. 16.32 a rendering of the 3D model of the house and its garden. Working to the given details construct and render the 3D model. (Fig. 16.33)

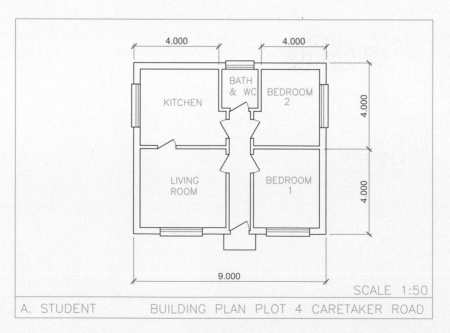

Fig. 16.31 Exercise 3

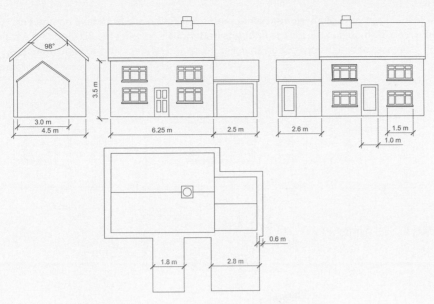

Fig. 16.32 Exercise 4 – orthographic views

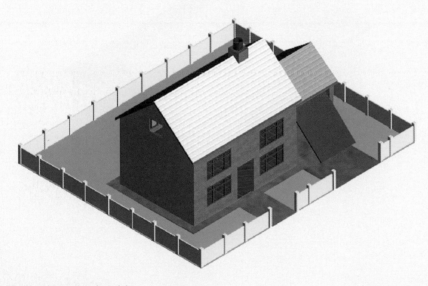

Fig. 16.33 Exercise 4 – the rendered model

5.	Fig. 16.34 is a three view, dimensioned orthographic projection of a brick built garage. Fig. 16.35 is a rendering of a 3D model of the garage minus its front door. Construct the 3D model to a scale of 1:50, making estimates of dimensions not given in Fig. 16.34 and render using suitable materials.

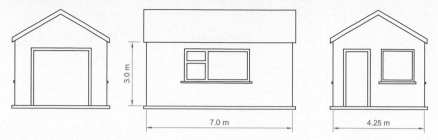

Fig. 16.34 Exercise 5 – orthographic views

Fig. 16.35 Exercise 5

Three-dimensional space

AIMS OF THIS CHAPTER

The aims of this chapter are:

1. to show in examples the methods of manipulating 3D models in 3D space using tools from the **View/UCS** panel or from the command line.
2. to introduce the Surfaces tools from the **Home/3D Modeling** panel

CHAPTER 17

3D space

So far in this book, when constructing 3D model drawings, they have been constructed on the AutoCAD 2009 coordinate system which is based upon three planes:

- the **XY Plane** – the screen of the computer;
- the **XZ Plane** at right angles to the **XY Plane** and as if coming towards the operator of the computer;
- a third plane (**YZ**) lying at right angles to both the other two planes (Fig. 17.1).

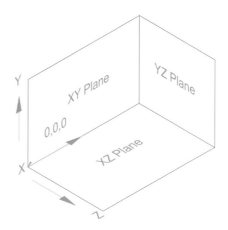

Fig. 17.1 The 3D space planes

In earlier chapters, the **ViewCube** was described, which enables 3D objects which have been constructed on these three planes to be viewed from different viewing positions. Other methods of rotating the model in 3D space and placing the model in other viewing positions using the **Vpoint Presets** dialog and the **3Dorbit** tool have also been described.

The **User Coordinate System (UCS)**

> **Note**
>
> The **XY** plane is the basic **UCS** plane, which in terms of the UCS is known as the ***WORLD*** plane.

The **UCS** allows the operator to place the AutoCAD coordinate system in any position in 3D space using a variety of **UCS** tools (commands). Features of the **UCS** can be called by *entering* **ucs** at the command line, or by selection of tools from the **UCS** panel (Fig. 17.2) or from the two **UCS** toolbars – **UCS** and **UCS II** (Fig. 17.3).

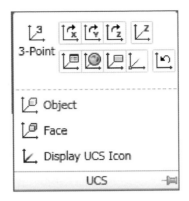

Fig. 17.2 The **View/UCS** panel

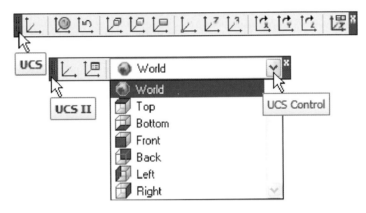

Fig. 17.3 The tools in the two **UCS** toolbars

If **ucs** is *entered* at the command line, it shows:

```
Command: enter ucs right-click
Current ucs name: *WORLD*
Specify origin of UCS or [Face/NAmed/OBject/
Previous/View/World/X/Y/Z/ZAxis] <World>:
```

and from these prompts selection can be made.

The variable UCSFOLLOW

UCS planes can be set from using any of the methods given above (Figs 17.2 and 17.3) or by *entering* **ucs** at the command line. No matter which method is used, the variable **UCSFOLLOW** must first be set on as follows:

```
Command: enter ucsfollow right-click
Enter new value for UCSFOLLOW <0>: enter 1
  right-click
Command:
```

The **UCS** icon

The **UCS** icon indicates the direction in which the three coordinate axes **X**, **Y** and **Z** lie in the AutoCAD drawing. When working in 2D, only the **X** and **Y** axes are showing, but when the drawing area is in a 3D view all three coordinate arrows are showing, except when the model is in the **XY** plane. The icon can be turned off as follows:

```
Command: enter ucsicon right-click
Enter an option [ON/OFF/All/Noorigin/ORigin/
  Properties] <ON>:
```

To turn the icon off, *enter* **off** in response to the prompt line and the icon disappears from the screen.

The appearance of the icon can be changed by *entering* **p** (Properties) in response to the prompt line. The **UCS Icon** dialog appears in which changes can be made to the shape, line width and colour of the icon if wished.

Types of **UCS** icon

The shape of the icon can be varied partly when changes are made in the **UCS Icon** dialog but also according to whether the AutoCAD drawing area is in 2D, 3D or **Paper Space** (Fig. 17.4).

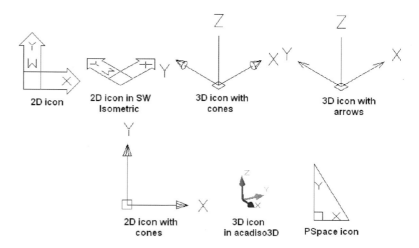

Fig. 17.4 Types of **UCS** icon

Examples of changing planes using the UCS

First example – changing UCS planes (Fig. 17.6)

1. Set UCSFOLLOW to 1 (ON).
2. Place the screen in **ViewCube/Front** and **Zoom** to **1**.

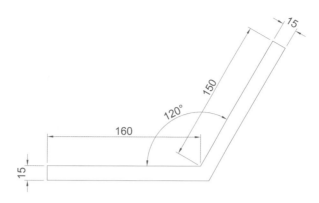

Fig. 17.5 First example – Changing UCS planes – pline for extrusion

3. Construct the pline outline in Fig. 17.5 and extrude to 120 high.
4. Place in **ViewCube/Isometric** view and **Zoom** to **1**.
5. With the **Fillet** tool, fillet corners to a radius of 20.
6. At the command line:

```
Command: enter ucs right-click
Current ucs name: *WORLD*
Specify origin of UCS or [Face/NAmed/OBject/
  Previous/View/World/X/Y/Z/ZAxis] <World>: enter
  f (Face) right-click
Select face of solid object: pick the sloping
  face - its outline highlights
Enter an option [Next/Xflip/Yflip] <accept>:
  right-click
Command:
```

And the 3D model changes its plane so that the sloping face is now on the new UCS plane. **Zoom** to **1**.

7. On this new UCS, construct four cylinders of radius 7.5 and height −15 (note the minus) and subtract them from the face.
8. *Enter* **ucs** at the command line again and *right-click* to place the model in the ***WORLD* UCS**.
9. Place four cylinders of the same radius and height into position in the base of the model and subtract them from the model.
10. Place the 3D model in **ViewCube/Isometric** and set in the **Home/View/Conceptual** visual style (Fig. 17.6).

Second example – UCS (Fig. 17.11)

The 3D model for this example is a steam venting valve – a two-view third-angle projection of the valve is shown in Fig. 17.7.

1. Make sure that **UCSFOLLOW** is set to **1**.
2. Place in the **UCS *WORLD** view. Construct the **120** square plate at the base of the central portion of the valve. Construct five cylinders for the holes in the plate. Subtract the five cylinders from the base plate.

Fig. 17.6 First example – Changing UCS planes

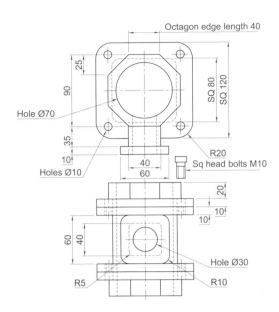

Fig. 17.7 Second example – UCS – the orthographic projection of a steam venting valve

3. Construct the central part of the valve – a filleted **80** square extrusion with a central hole.
4. At the command line:

```
Command: enter ucs right-click
Current ucs name: WORLD*
Specify origin of UCS or [Face/NAmed/OBject/
  Previous/View/World/X/Y/Z/ZAxis] <World>:
  enter x right-click
Specify rotation angle about X axis <90>:
  right-click
Command:
```

Fig. 17.8 Second example – UCS – step 11 + rendering

and the model assumes a **Front** view.

5. With the **Move** tool, move the central portion vertically up by **10**.
6. With the **Copy** tool, copy the base up to the top of the central portion.
7. With the **Union** tool form a single 3D model of the three parts.
8. Make the layer **Construction** current.
9. Place the model in the **UCS *WORLD*** view. Construct the separate top part of the valve – a plate forming a union with a hexagonal plate and with holes matching those of the other parts.
10. Place the drawing in the **UCS X** view. Move the parts of the top into their correct positions relative to each other. With **Union** and **Subtract** complete the part. This will be made easier if the layer **0** is turned off.
11. Turn layer **0** back on and move the top into its correct position relative to the main part of the valve. Then with the **Mirror** tool, mirror the top to produce the bottom of the assembly.

Fig. 17.9 Second example – UCS – steps 12 and 13 + rendering

Fig. 17.10 Second example – UCS – pline for the bolt

Fig. 17.11 Second example – UCS

12. While in the **UCS X** view construct the three parts of a 3D model of the extrusion to the main body.
13. In the **UCS *WORLD*** view, move the parts into their correct position relative to each other. **Union** the two filleted rectangular extrusions and the main body. **Subtract** the cylinder from the whole.
14. In the **UCS X** view, construct one of the bolts as shown in Fig. 17.10, forming a solid of revolution from a pline. Then construct a head to the bolt and with **Union** add it to the screw.
15. With the **Copy** tool, copy the bolt 7 times to give 8 bolts. With **Move**, and working in the **UCS *WORLD*** and **X** views, move the bolts into their correct positions relative to the 3D model.
16. Add suitable lighting and attach materials to all parts of the assembly and render the model.
17. Place the model in the **ViewCube/Isometric** view.
18. Save the model to a suitable file name.
19. Finally, move all the parts away from each other to form an exploded view of the assembly (Fig. 17.11).

Third example – UCS (Fig. 17.15)

1. Set **UCSFOLLOW** to **1**.
2. Place the drawing area in the **UCS X** view.
3. Construct the outline in Fig. 17.12 and extrude to a height of 120.
4. *Click* the **UCS, 3 Point** tool icon in the **View/UCS** panel (Fig. 17.13):

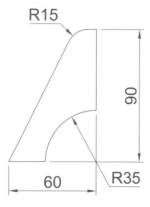

Fig. 17.12 Third example – UCS – outline for 3D model

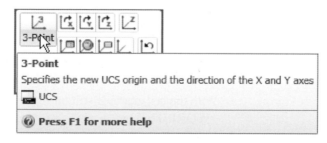

Fig. 17.13 The **UCS, 3 Point** icon in the **View/UCS** panel

```
Command: _ucs
Current ucs name: *WORLD*
Specify origin of UCS or [Face/NAmed/OBject/
  Previous/View/World/X/Y/Z/ZAxis] <World>: _3
Specify new origin point <0,0,0>: pick point
  (Fig. 17.14)
Specify point on positive portion of X-axis: pick
  point (Fig. 17.14)
Specify point on positive-Y portion of the UCS XY
  plane <-142,200,0>: enter .xy right-click
of (need Z): enter 1 right-click
Regenerating model
Command:
```

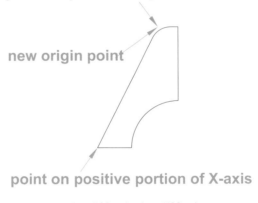

point on positive -Y portion of the UCS XY plane

new origin point

point on positive portion of X-axis

Fig. 17.14 Third example – UCS – the three UCS points

Figure 17.14 shows the **UCS** points and the model regenerates in this new 3 point plane.

5. On the face of the model construct a rectangle **80 × 50** central to the face of the front of the model, fillet its corners to a radius of 10 and extrude to a height of 10.
6. Place the model in the **ViewCube/Isometric** view and fillet the back edge of the second extrusion to a radius of 10.
7. Subtract the second extrusion from the first.
8. Add lights, and a suitable material, and render the model (Fig. 17.15).

Fourth example – UCS (Fig. 17.17)

1. With the last example still on screen, place the model in the **UCS *WORLD*** view.
2. Call the **Rotate** tool from the **Home/Modify** panel and rotate the model through 225 degrees.

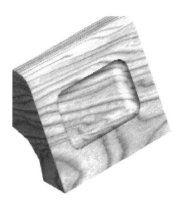

Fig. 17.15 Third example – UCS

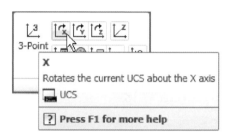

Fig. 17.16 The **UCS X** tool in the **View/UCS** panel

3. *Click* the **UCS, X Axis Vector** tool icon in the **UCS** toolbar (Fig. 17.16):

```
Command: _ucs
Current ucs name: *WORLD*
Specify origin of UCS or [Face/NAmed/OBject/
  Previous/View/World/X/Y/Z/ZAxis] <World>: _x
Specify rotation angle about X axis <90>:
  right-click
Regenerating model.
Command:
```

4. Render the model in its new **UCS** plane (Fig. 17.17)

Fig. 17.17 Fourth example

Saving UCS views

If a number of different UCS planes are used in connection with the construction of a 3D model, each can be saved to a different name and recalled when required. To save the UCS plane in which a 3D model drawing is being constructed *enter* **ucs** at the command line:

```
Current ucs name: *NO NAME*
Specify origin of UCS or [Face/NAmed/OBject/
  Previous/View/World/X/Y/Z/ZAxis] <World>:
  enter s right-click
Enter name to save current UCS or [?]: enter New
  View right-click
Regenerating model
Command:
```

Now *click* the **UCS Control** in the **UCS II** toolbar and the list of **UCS** views appears showing the names of views saved in the current drawing (Fig. 17.18).

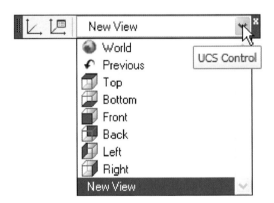

Fig. 17.18 The **UCS Control** list in the **UCS II** toolbar

Constructing 2D objects in 3D space

In previous chapters of this book there have been examples of 2D objects constructed with the **Polyline**, **Line**, **Circle** and other 2D tools to form the outlines for extrusions and solids of revolution. These outlines have been drawn on planes in the **ViewCube** settings.

First example – 2D outlines in 3D space (Fig. 17.21)

1. Construct a **3point UCS** to the following points:

```
Origin point: 80,90
X-axis point: 290,150
```

```
Positive-Y point: .xy of 80,90
(need Z): enter 1.
```

2. On this **3point UCS** construct a 2D drawing of the plate to the dimensions given in Fig. 17.19, using the **Polyline**, **Ellipse** and **Circle** tools.

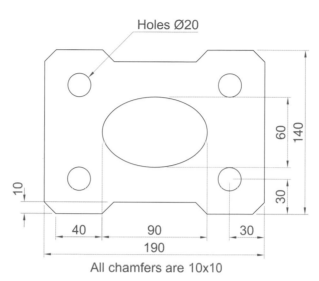

Holes Ø20

All chamfers are 10x10

Fig. 17.19 First example – 2D outlines in 3D space

3. Save the **UCS** plane in the **UCS** dialog to the name **3point**.
4. Place the drawing area in the **ViewCube/Isometric** view (Fig. 17.20).
5. With the **Region** tool form regions of the 6 parts of the drawing and with the **Subtract** tool, subtract the circles and ellipse from the main outline.
6. Place in the **Home/View/Conceptual** visual style. Extrude the region to a height of 10 (Fig. 17.21).

Fig. 17.20 First example – 2D outlines in 3D space. The outline in the **Isometric** view

Fig. 17.21 First example – 2D outlines in 3D space

Second example – 2D outlines in 3D space (Fig. 17.25)

1. Place the drawing area in the **ViewCube/Front** view, **Zoom** to **1** and construct the outline in Fig. 17.22.
2. Extrude the outline to 150 high.

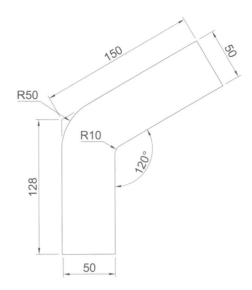

Fig. 17.22 Second example – 2D outlines in 3D space. Outline to be extruded

3. Place in the **ViewCube/Isometric** view and **Zoom** to **1**.
4. *Click* the **Face** tool icon in the **View/UCS** toolbar (Fig. 17.23) and place the 3D model in the **UCS** plane shown in Fig. 17.24, selecting the sloping face of the extrusion for the plane and again **Zoom** to **1**.
5. With the **Circle** tool draw five circles as shown in Fig. 17.24.

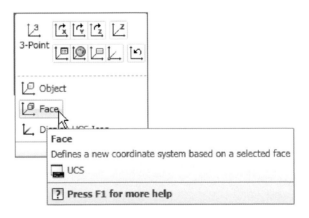

Fig. 17.23 The **Face** icon form the **View/UCS** toolbar

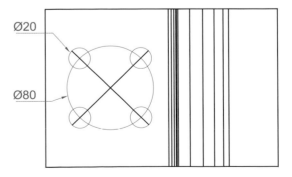

Fig. 17.24 Second example – 2D outlines in 3D space – the circles in the new **UCS FACE**

Fig. 17.25 Second example – 2D outlines in 3D space

6. Form a region from the five circles and with **Union** form a union of the regions.
7. Extrude the region to a height of –60 (Note the minus) – higher than the width of the sloping part of the 3D model.
8. Place the model in the **ViewCube/Isometric** view and subtract the extruded region from the model.
9. With the **Fillet** tool, fillet the upper corners of the slope of the main extrusion to a radius of 30.
10. Place the model into another **UCS FACE** plane and construct a filleted pline of sides 80 and 50 and filleted to a radius of 20. Extrude to a height of –60 and subtract the extrusion from the 3D model.
11. Place in the **ViewCube/Isometric** view; add lighting and a material.

The result is shown in Fig. 17.25.

The Surfaces tools

The construction of 3D surfaces from lines, arc and plines has been dealt with on pages 228–229. In this section, examples of 3D surfaces constructed with the tools **Edgesurf**, **Rulesurf** and **Tabsurf** will be described. The tools can be called from the **Home/3D Modeling** panel. Figure 17.26 shows the **Tabulated Surface** tool icon in the panel. The two icons to the right of that shown are the **Ruled Surface** and the **Edge Surface** tools. In this section these three surface tools will be called by *entering* their tool names at the command line.

Surface meshes

Surface meshes are controlled by the set variables **Surftab1** and **Surftab2**. These variables are set as follows:

At the command line:

```
Command: enter surftab1 right-click
Enter new value for SURFTAB1 <6>: enter 24
  right-click
Command:
```

Fig. 17.26 The **Tabulated Surface** tool icon in the **Home/3D Modeling** panel

The Edgesurf tool – Fig. 17.29

1. Make a new layer colour **magenta**. Make that layer current.
2. Place the drawing area in the **View Cube/Right** view. **Zoom** to **All**.
3. Construct the polyline to the sizes and shape as shown in Fig. 17.27.
4. Place the drawing area in the **View Cube/Top** view. **Zoom** to **All**.
5. Copy the pline to the right by 250.
6. Place the drawing in the **View Cube/Isometric** view. **Zoom** to **All**.
7. With the **Line** tool, draw lines between the ends of the two plines using the **endpoint** osnap (Fig. 17.28). Note that if polylines are drawn they will not be accurate at this stage.

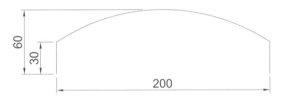

Fig. 17.27 Example – **Edgesurf** – pline outline

Fig. 17.28 Example – **Edgesurf** – adding lines joining the plines

8. Set **Surftab1** to **32** and **Surftab2** to **64**.
9. At the command line:

```
Command: enter edgesurf right-click
Current wire frame density: SURFTAB1=32
  SURFTAB2=64
Select object 1 for surface edge: pick one of the
  lines (or plines)
Select object 2 for surface edge: pick the next
  adjacent line (or pline)
Select object 3 for surface edge: pick the next
  adjacent line (or pline)
Select object 4 for surface edge: pick the last
  line (or pline)
Command:
```

The result is shown in Fig. 17.29.

The Rulesurf tool – Fig. 17.31

1. Make a new layer colour **blue** and make the layer current.
2. In the **3D Navigate/Front** view construct the pline as shown in Fig. 17.30.

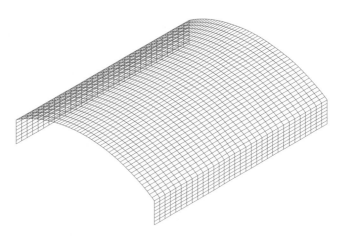

Fig. 17.29 Example – **Edgesurf**

Fig. 17.30 **Rulesurf** – the outline

Fig. 17.31 Example – **Rulesurf**

3. In the **3D Navigate/Top**, **Zoom** to **1** and copy the pline to a vertical distance of 120.
4. Place in the **3D Navigate/Southwest Isometric** view and **Zoom** to **1**.
5. Set **SURFTAB1** to **32**.
6. At the command line:

```
Command: enter rulesurf right-click
Current wire frame density: SURFTAB1 = 32
Select first defining curve: pick one of the
  plines
Select second defining curve: pick the other pline
Command:
```

The result is given in Fig. 17.31.

The Tabsurf tool – Fig. 17.32

1. Make a new layer of colour blue and make the layer current.
2. Set **Surftab1** to **2**.
3. In the **3D Navigate/Top view** construct a hexagon of edge length 35.
4. In the **3D Navigate/Front** view and in the centre of the hexagon construct a pline of height 100.
5. Place the drawing in the **3D Navigate/Southwest Isometric** view.
6. At the command line:

```
Command: enter tabsurf right-click
Current wire frame density: SURFTAB1=2
Select objects for path curve: pick the hexagon
Select object for direction vector: pick the pline
Command:
```

See Fig. 17.32.

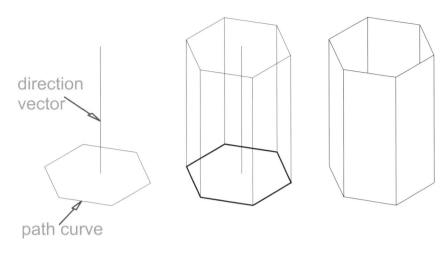

direction
vector

path curve

Fig. 17.32 Example – **Tabsurf**

REVISION NOTES

1. The **UCS** (User Coordinate System) tools can be called from the **View/UCS** panel, from the two toolbars **UCS** and **UCS II** or by *entering* **ucs** at the command line.
2. The variable **UCSFOLLOW** must first be set on (to **1**) before operations of the UCS can be brought into action.
3. There are several types of **UCS** icon – **2D** different types, **3D** (different types), **Pspace.**
4. The position of the plane in 3D space on which a drawing is being constructed can be varied using tools from the **UCS** panel.
5. The planes on which drawings constructed on different planes in 3D space can be saved in the **UCS II** view menu.
6. The tools **Edgesurf, Rulesurf** and **Tabsurf** can be used to construct surfaces in addition to surfaces which can be constructed from plines and lines using the **Extrude** tool.

Exercises

Methods of constructing answers to the following exercises can be found in the free website:

http://www.books.elsevier.com/companions/9780750689830

1. Figure 17.33 shows a rendering of a two-view projection of an angle bracket in which two pins are placed in holes in each of the arms of the bracket. Figure 17.34 shows a two-view projection of the bracket. Construct a 3D model of the bracket and its pins. Add lighting to the scene and materials to the parts of the model and render.

Fig. 17.33 Exercise 1 – a rendering

2. The two-view projection (Fig. 17.35) shows a stand consisting of two hexagonal prisms. Circular holes have been cut right through each face of the smaller hexagonal prism and rectangular holes with rounded ends have been cut right through the faces of the larger. Construct a 3D model of the stand. When completed add suitable lighting to the scene. Then add a material to the model and render (Fig. 17.36).

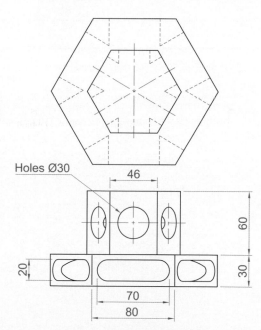

Fig. 17.35 Exercise 2 – details of shapes and sizes

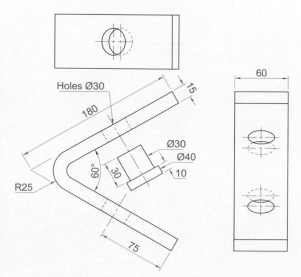

Fig. 17.34 Exercise 1 – details of shapes and sizes

Fig. 17.36 Exercise 2 – a rendering

3.   The two-view projection (Fig. 17.37) shows a ducting pipe. Construct a 3D model drawing of the pipe. Place in a **SW Isometric** view; add lighting to the scene and a material to the model; and render.

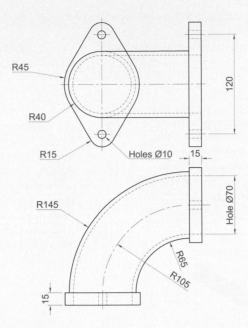

Fig. 17.37 Exercise 3 – details of shapes and sizes

4.   A point marking device is shown in two two-view projections (Fig. 17.38). The device is composed of three parts – a base, an arm and a pin. Construct a 3D model of the assembled device and add

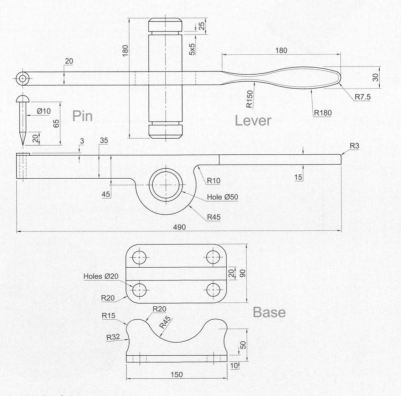

Fig. 17.38 Exercise 4 – details of shapes and sizes

appropriate materials to each part. Then add lighting to the scene and render in an **SW Isometric** view (Fig. 17.39).

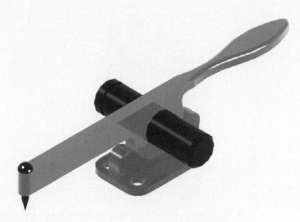

Fig. 17.39 Exercise 4 – a rendering

5.   A rendering of a 3D model drawing of the connecting device shown in the orthographic projection shown in Fig. 17.41 is given in Fig. 17.40. Construct the 3D model drawing of the device and add a suitable lighting to the scene. Then place in the **ViewCube/Isometric** view, add a material to the model and render.

Fig. 17.40 Exercise 5 – a rendering

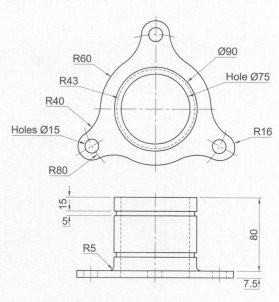

R60

Ø90

R43

Hole Ø75

R40

Holes Ø15

R16

R80

15

5

R5

80

7.5

Fig. 17.41 Exercise 5 – two-view drawing

6.   A fork connector and its rod are shown in a two-view projection in Fig. 17.42. Construct a 3D model drawing of the connector with its rod in position. Then add lighting to the scene, place in the **ViewCube/Isometric** viewing position, add materials to the model and render.

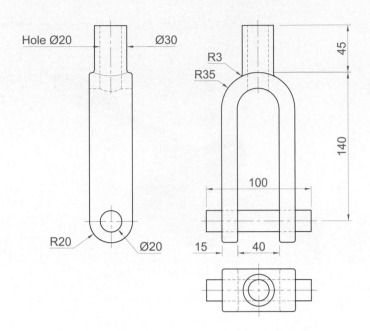

Fig. 17.42 Exercise 6

7. An orthographic projection of the parts of a lathe steady are given in Fig. 17.43. From the dimensions shown in the drawing, construct an assembled 3D model of the lathe steady. When the 3D model has been completed, add suitable lighting and materials and render the model (Fig. 17.44).

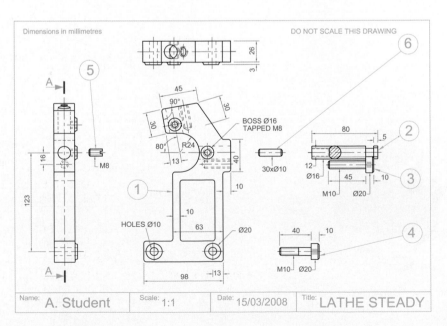

Fig. 17.43 Exercise 7 – details

Fig. 17.44 Exercise 7 – a rendering

8. Construct suitable polylines to sizes of your own discretion in order to form the two surfaces to form the box shape shown in Fig. 17.45 with the aid of the **Rulesurf** tool. Add lighting and a material and render the surfaces so formed. Construct another three **Edgesurf** surfaces to form a lid for the box. Place the surface in a position above the box, add a material and render (Fig. 17.46).

Fig. 17.45 Exercise 8 – the box

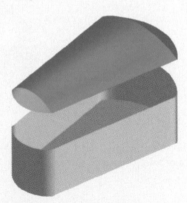

Fig. 17.46 Exercise 8 – the box and its lid

9. Figure 17.47 shows a polyline for each of the 4 objects from which the surface shown in Fig. 17.48 was obtained. Construct the surface and shade with **Realistic** shading.

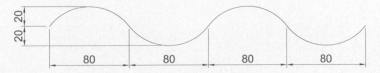

Fig. 17.47 Exercise 9 – one of the solutions from which the surface was obtained

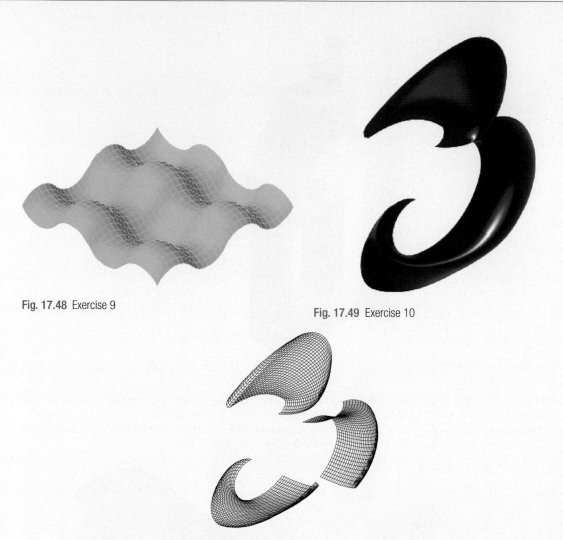

Fig. 17.48 Exercise 9

Fig. 17.49 Exercise 10

Fig. 17.50 The three surfaces

10. The surface model for this exercise was constructed from 3 **Edgesurf** surfaces working to the suggested objects for the surface as shown in Fig. 17.51. The sizes of the outlines of the objects in each case are

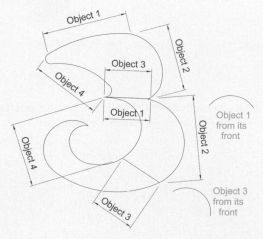

Fig. 17.51 Outlines of the three surfaces

left to your discretion. Figure 17.49 shows the completed surface model. Figure 17.50 shows the three surfaces of the model separated from each other.

11.   Figure 17.52 shows in a **View Block/Isometric** view a semicircle of radius 25 constructed in the **View Cube/Top** view on a layer of colour **Magenta** with a semicircle of radius 75 constructed on the **View Block/Front** view with its left-hand end centred on the semicircle. Figure 17.53 shows a surface constructed from the two semicircles in a **Visual Styles/Realistic** mode.

Fig. 17.52 Exercise 11 – the circle and semicircle

Fig. 17.53 Exercise 11

Chapter 18

Editing 3D solid models

The aims of this chapter are:

1. to introduce the use of tools from the **Solid Editing** panel;
2. to show examples of a variety of 3D solid models.

The Solid Editing tools

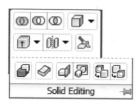

Fig. 18.1 The **Home/Solid Editing** panel

The **Solid Editing** tools can be called from the **Home/Solid Editing** panel (Fig. 18.1) or from the **Solid Editing** toolbar (Fig. 18.2).

Examples of the results of using some of the **Solid Editing** tools are shown in this chapter. These tools are of value if the design of a 3D solid model needs to be changed (edited), although some are useful for constructing parts of 3D solids which cannot easily be constructed using other tools.

Fig. 18.2 The **Solid Editing** toolbar

First example – **Extrude faces** tool (Fig. 18.5)

1. Set **ISOLINES** to 24.
2. In the **ViewCube/Right** construct a cylinder of radius 30 and height 30 (Fig. 18.3).
3. In the **ViewCube/Front** construct the pline shown in Fig. 18.3. Mirror the pline to the other end of the cylinder.

Fig. 18.4 The **Extrude faces** tool from the **Home/Solid Editing** panel

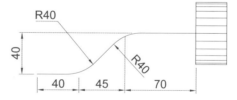

Fig. 18.3 First example – **Extrude faces** tool – first stages

4. In the **ViewCube/Top** move the pline to lie central to the cylinder.
5. Place the screen in the **ViewCube/Isometric**.
6. *Click* the **Extrude faces** tool icon in the **Home/Solid Editing** panel (Fig. 18.4). The command line shows:

```
Command: _solidedit
Solids editing automatic checking: SOLIDCHECK=1
Enter a solids editing option [Face/Edge/Body/
  Undo/eXit] <eXit>: _face
Enter a face editing option
[Extrude/Move/Rotate/Offset/Taper/Delete/Copy/
  coLor/mAterial/Undo/eXit] <eXit>: _extrude
Select faces or [Undo/Remove]: 2 faces found.
Select faces or [Undo/Remove/ALL]: enter r
  right-click
```

Original
cylinder

After extruding
faces along paths

Fig. 18.5 First example –
Extrude faces tool

```
Remove faces or [Undo/Add/ALL]: pick 1 face found,
  1 removed.
Specify height of extrusion or [Path]: enter p
  right-click
Select extrusion path: pick the path pline
Solid validation started.
Solid validation completed.
Enter a face editing option [Extrude/Move/Rotate/
  Offset/Taper/Delete/Copy/coLor/mAterial/Undo/
  eXit] <eXit>: right-click
Command:
```

7. Repeat the operation using the view at the other end of the cylinder.
8. Add lights and a material and render the 3D model (Fig. 18.5).

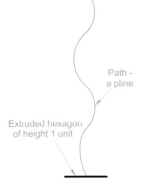

Path –
a pline

Extruded hexagon
of height 1 unit

Fig. 18.6 Second example –
Extrude faces tool – pline
for path

> **Note**
>
> Note the prompt line which includes the statement
> **SOLIDCHECK = 1**. If the variable **SOLIDCHECK** is set on (to **1**)
> the prompt lines include the lines **SOLIDCHECK = 1**, **Solid
> validation started** and **Solid validation completed**. If set to **0** these
> two lines do not show.

Second example – **Extrude faces** tool (Fig. 18.7)

1. Construct a hexagonal extrusion just **1** unit high in the **ViewCube/Top**.
2. Change to the **ViewCube/Front** and construct the curved pline (Fig. 18.6).
3. Back in the **Top** view, move the pline to lie central to the extrusion.
4. Place in the **ViewCube/Isometric** and extrude the top face of the extrusion along the path of the curved pline.
5. Add lighting and a material to the model and render (Fig. 18.7).

> **Note**
>
> This example shows that a face of a 3D solid model can be extruded
> along any suitable path curve. If the polygon on which the extrusion had
> been based had been turned into a region, no extrusion could have taken
> place. The polygon had to be extruded to give a face to a 3D solid.

Fig. 18.7 Second example –
Extrude faces tool

Third example – **Move faces** tool (Fig. 18.8)

1. Construct the 3D solid drawing shown in the left-hand drawing of Fig. 18.8 from three boxes which have been united using the **Union** tool.

2. *Click* on the **Move faces** tool in the **Home/Solid Editing** panel (see Fig. 18.4, page 344). The command line shows:

```
Command: _solidedit
[prompts]: _face
Enter a face editing option
[prompts]: _move
Select faces or [Undo/Remove]: pick face 1 face
   found.
Select faces or [Undo/Remove/ALL]: right-click
Specify a base point or displacement: pick
Specify a second point of displacement: pick
[further prompts]:
```

and the *picked* face is moved – right-hand drawing of Fig. 18.8.

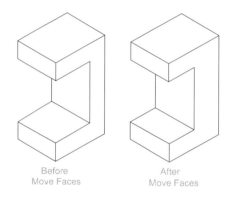

Before
Move Faces

After
Move Faces

Fig. 18.8 Third example – **Solid, Move faces** tool

Fourth example – **Offset faces** tool (Fig. 18.9)

1. Construct the 3D solid drawing shown in the left-hand drawing of Fig. 18.9 from a hexagonal extrusion and a cylinder which have been united using the **Union** tool.
2. *Click* on the **Offset faces** tool icon in the **Home/Solid Editing** toolbar (see Fig. 18.4). The command line shows:

```
Command: _solidedit
[prompts]: _face
[prompts]
[prompts]: _offset
Select faces or [Undo/Remove]: pick the bottom
   face of the 3D model 2 faces found.
Select faces or [Undo/Remove/All]: enter r
   right-click
Select faces or [Undo/Remove/All]: pick
   highlighted faces other than the bottom face 2
   faces found, 1 removed
```

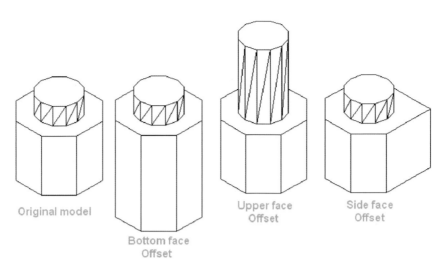

Fig. 18.9 Third example – **Offset faces** tool

```
Select faces or [Undo/Remove/All]: right-click
Specify the offset distance: enter 30 right-click
```

3. Repeat, offsetting the upper face of the cylinder by **50** and the right-hand face of the lower extrusion by **15**.

    The results are shown in Fig. 18.9.

Fifth example – **Taper faces** tool (Fig. 18.10)

1. Construct the 3D model as in the left-hand drawing of Fig. 18.10. Place in **ViewCube/Isometric**.
2. Call **Taper faces**. The command line shows:

```
Command: _solidedit
[prompts]: _face
[prompts]
[prompts]: _taper
Select faces or [Undo/Remove]: pick the upper face
  of the base 2 faces found.
Select faces or [Undo/Remove/All]: enter r
  right-click
```

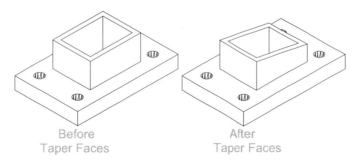

Before
Taper Faces

After
Taper Faces

Fig. 18.10 Fifth example – **Taper faces** tool

```
Select faces or [Undo/Remove/All]: pick
  highlighted faces other than the upper face 2
  faces found, 1 removed
Select faces or [Undo/Remove/All]: right-click
Specify the base point: pick a point on the
  right-hand edge of the face
Specify the taper angle: enter 10 right-click
```

and the selected face tapers as indicated in the right-hand drawing (Fig. 18.10).

Sixth example – **Copy faces** tool (Fig. 18.12)

1. Construct a 3D model to the sizes as given in Fig. 18.11.

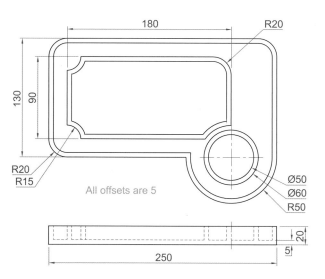

Fig. 18.11 Sixth example – **Copy faces** tool – details of the 3D solid model

2. *Click* on the **Copy faces** tool in the **Home/Solid Editing** toolbar (see Fig. 18.4, page 344). The command line shows:

```
Command: _solidedit
[prompts]: _face
[prompts]
[prompts]: _copy
Select faces or [Undo/Remove]: pick the upper face
  of the solid model 2 faces found.
Select faces or [Undo/Remove/All]: enter r
  right-click
Select faces or [Undo/Remove/All]: pick
  highlighted face not to be copied 2 faces found,
  1 removed
Select faces or [Undo/Remove/All]: right-click
```

```
Specify a base point or displacement:
  pick  anywhere on the highlighted face
Specify a second point of displacement: pick a
  point some 50 units above the face
```

3. Add lights and a material to the 3D model and its copied face and render (Fig. 18.12).

Fig. 18.12 Sixth example –
Copy faces tool

Seventh example – **Color faces** tool (Fig. 18.14)

1. Construct a 3D model of the wheel to the sizes as shown in Fig. 18.13.

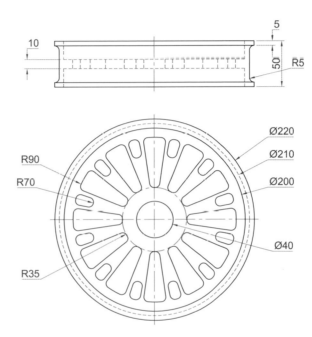

Fig. 18.13 Seventh example – **Color faces** tool – details of the 3D model

2. *Click* the **Color faces** tool icon in the **Home/Solid Editing** toolbar (see Fig. 18.4, page 344). The command line shows:

```
Command: _solidedit
[prompts]: _face
[prompts]
[prompts]: _color
Select faces or [Undo/Remove]: pick the inner face
  of the wheel 2 faces found
Select faces or [Undo/Remove/All]: enter r
  right-click
Select faces or [Undo/Remove/All]: pick
  highlighted faces other than the required face 2
  faces found, 1 removed
Enter new color <ByLayer>: enter 1 (which is red)
  right-click
```

3. Add lights and a material to the edited 3D model and render (Fig. 18.14).

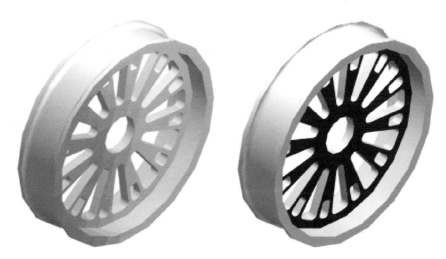

Fig. 18.14 Seventh example – **Color faces** tool

Examples of more 3D models

These 3D models can be constructed in the **acadiso3D.dwt** screen. The descriptions of the stages needed to construct these 3D models have been reduced from those given in earlier pages, in the hope that readers have already acquired a reasonable skill in the construction of such drawings.

First example (Fig. 18.16)

1. **Front** view. Construct the three extrusions for the back panel and the two extruding panels to the details given in Fig. 18.15.

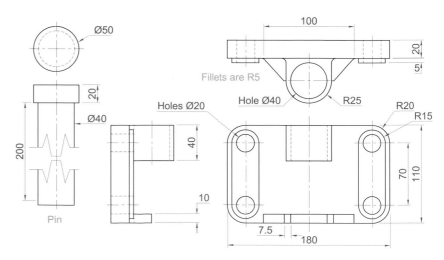

Fig. 18.15 First example – **3D models** – details of sizes and shapes

Fig. 18.16 First example – **3D models**

2. **Top** view. Move the two panels to the front of the body and union the three extrusions. Construct the extrusions for the projecting parts holding the pin.
3. **Front** view. Move the two extrusions into position and union them to the back.
4. **Top** view. Construct two cylinders for the pin and its head.
5. **Top** view. Move the head to the pin and union the two cylinders.
6. **Front** view. Move the pin into its position in the holder. Add lights and materials.
7. **Isometric** view. Render. Adjust lighting and materials as necessary (Fig. 18.16).

Second example (Fig. 18.18)

1. **Top** (Fig. 18.17). Construct polyline outlines for the body extrusion and the solids of revolution for the two end parts. Extrude the body and subtract its hole and using the **Revolve** tool form the two end solids of revolution.

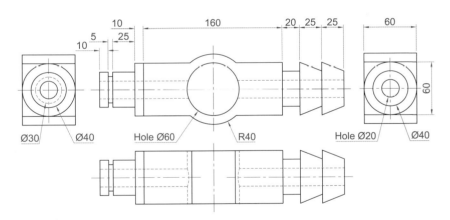

Fig. 18.17 Second example – 3D models dimensions

2. **Right**. Move the two solids of revolution into their correct positions relative to the body and union the three parts. Construct a cylinder for the hole through the model.
3. **Front**. Move the cylinder to its correct position and subtract from the model.
4. **Top**. Add lighting and a material.
5. **Isometric**. Render (Fig. 18.18).

Third example (Fig. 18.18)

1. **Front**. Construct the three plines needed for the extrusions of each part of the model (details, Fig. 18.19). Extrude to the given heights. Subtract the hole from the **20** high extrusion.

Fig. 18.18 Second example – **3D models**

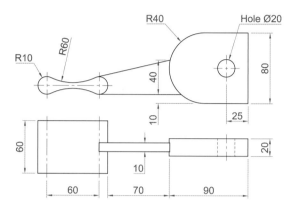

Fig. 18.19 Third example – **3D models** – details of shapes and sizes

Fig. 18.20 Third example – **3D models**

2. **Top**. Move the **60** extrusion and the **10** extrusion into their correct positions relative to the **20** extrusion. With **Union** form a single 3D model from the three extrusions.
3. Add suitable lighting and a material to the model.
4. **Isometric**. Render (Fig. 18.20).

Fourth example (Fig. 18.20)

1. **Front**. Construct the polyline – left-hand drawing in Fig. 18.21.

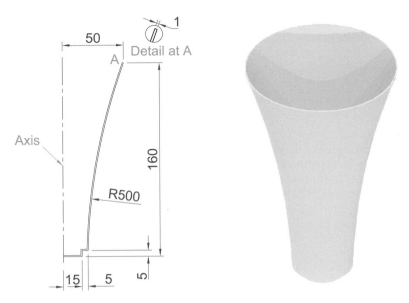

Fig. 18.21 Fourth example – **3D models**

2. With the **Revolve** tool from the **Home/3D modeling** panel construct a solid of revolution from the pline.
3. **Top**. Add suitable lighting and a coloured glass material.
4. **Isometric**. Render – right-hand illustration in Fig. 18.21.

Exercises

Methods of constructing answers to the following exercises can be found in the free website:

http://books.elsevier.com/companions/9780750689830

1. Working to the shapes and dimensions as given in the orthographic projection of Fig. 18.22, construct the exploded 3D model as shown in Fig. 18.23. When the model has been constructed add suitable lighting and apply materials, followed by rendering.

Add suitable lighting and materials, place in one of the isometric viewing positions and render the model.

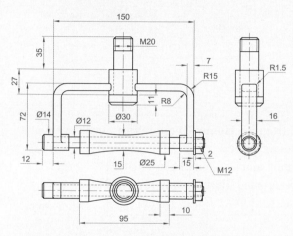

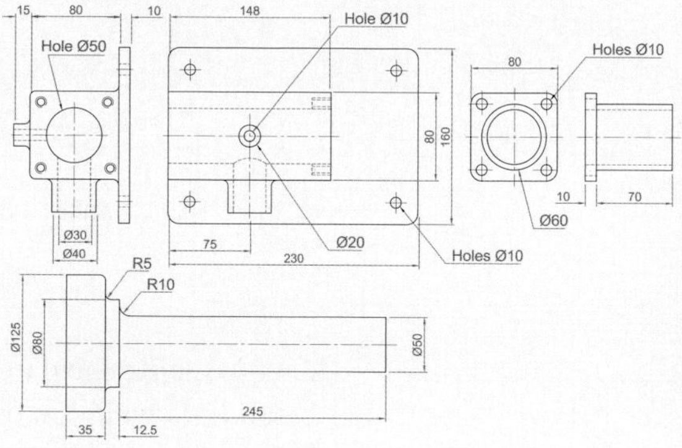

Fig. 18.24 Exercise 2 – details of shapes and sizes

Fig. 18.22 Exercise 1 – orthographic projection

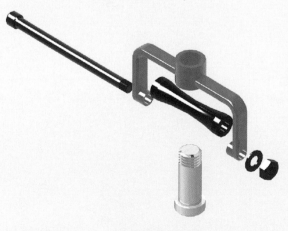

Fig. 18.25 Exercise 2

Fig. 18.23 Exercise 1 – rendered 3D model

3. Construct the 3D model shown in the rendering (Fig. 18.26) from the details given in the parts drawing of Fig. 18.27.

2. Working to the dimensions given in the orthographic projections of the three parts of this 3D model (Fig. 18.24), construct the assembled parts as shown in the rendered 3D model of Fig. 18.25.

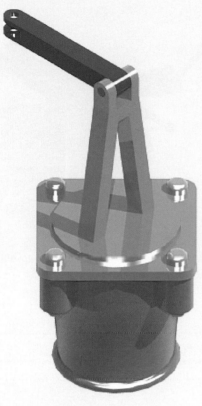

Fig. 18.26 Exercise 3

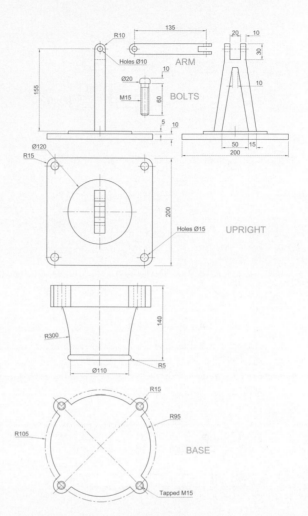

Fig. 18.27 Exercise 3 – the parts drawing

4. A more difficult exercise: a rendered 3D model of the parts of an assembly are shown in Fig. 18.31. Working to the details given in the three orthographic projections (Figs. 18.28, 18.29 and 18.30), construct the two parts of the 3D model, place them in suitable positions relative to each other, add lighting and materials and render the model.

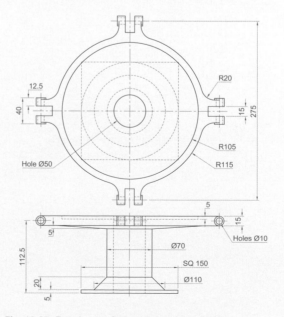

Fig. 18.28 Exercise 4 – first orthographic projection

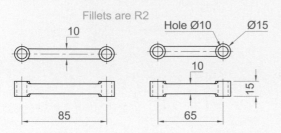

Fig. 18.29 Exercise 4 – second orthographic projection

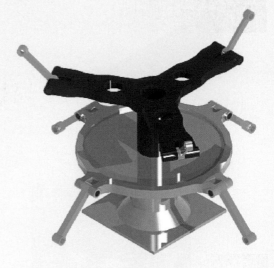

Fig. 18.31 Exercise 4

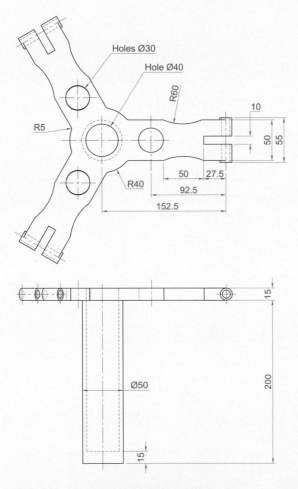

Fig. 18.30 Exercise 4 – third orthographic projection

Other features of 3D modelling

AIMS OF THIS CHAPTER

The aims of this chapter are:

1. to give a further example of placing raster images in an AutoCAD drawing;
2. to give examples of methods of printing or plotting not given in previous chapters;
3. to give examples of polygonal viewports.

Raster images in AutoCAD drawings

Example – Raster image in a drawing (Fig. 19.5)

This example shows the raster file **Fig14.bmp** of the 3D model constructed to the details given in the drawing of Fig. 19.1.

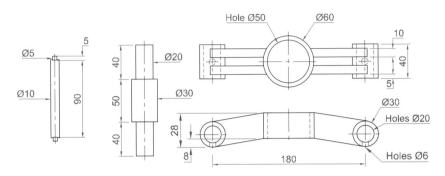

Fig. 19.1 Example – Raster image in a drawing – details

Raster images are graphics images in files with file names ending with the extensions ***.bmp**; ***.pcx**; ***.tif**; and the like. The types of graphics files which can be inserted into AutoCAD drawings can be seen by first *clicking* on the **Image** icon in the **Blocks & References/Reference** panel (Fig. 19.2), which brings the **Select Image File** dialog (Fig. 19.3) on screen. In the dialog *click* the arrow to the right of the **Files of type** field and the pop-up list which appears lists the types of graphics files which can be inserted into AutoCAD drawings. Such graphics files can be used to describe in 3D the details shown in a 2D technical drawing.

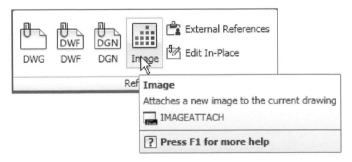

Fig. 19.2 Selecting **Image** from the **Blocks & References/References** panel

1. Construct the 3D model to the shapes and sizes given in Fig. 19.1 working in four layers, each of a different colour.
2. Place in the **ViewCube/Isometric** view.
3. Shade the 3D model in **Realistic** visual style.
4. **Zoom** the shaded model to a suitable size and press the **Print Scr** key of the keyboard.

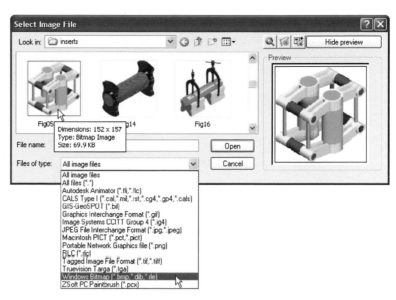

Fig. 19.3 The **Select Image File** dialog

5. Open the Windows **Paint** application and *click* **Edit** in the menu bar, followed by another *click* on **Paste** in the drop-down menu. The whole AutoCAD screen, which includes the shaded 3D assembled model, appears.
6. *Click* the **Select** tool icon in the toolbar of **Paint** and window the 3D model. Then *click* **Copy** in the **Edit** drop-down menu.
7. *Click* **New** in the **File** drop-down menu, followed by a *click* on **No** in the warning window which appears.
8. *Click* **Paste** in the **Edit** drop-down menu. The shaded 3D model appears. *Click* **Save As...** from the **File** drop-down menu and save the bitmap to a suitable file name – in this example, **Fig14.bmp**.

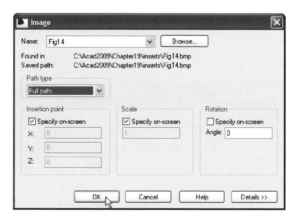

Fig. 19.4 Example – Raster image in a drawing – the **Image** dialog

9. Open the orthographic projection drawing of Fig. 19.1 in AutoCAD.
10. Open the **Select Image File** dialog and from the **Look in** field select the raster file **Fig14.bmp** from the file list (Fig. 19.3). Another dialog (**Image**) opens (Fig. 19.4) showing the name of the raster image file.

Click the **OK** button of the dialog and a series of prompts appears at the command line requesting position and scale of the image. *Enter* appropriate responses to these prompts and the image appears in position in the orthographic drawing (Fig. 19.5).

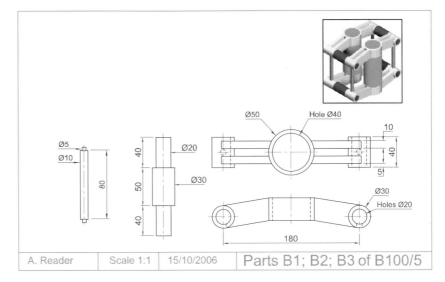

Fig. 19.5 Example – Raster image in a drawing

> **Note**
>
> 1. It will normally be necessary to *enter* a scale in response to the prompt lines otherwise the raster image may appear very small on screen. If it does it can be zoomed anyway.
>
> 2. Place the image in position in the drawing area. In Fig. 19.5 the orthographic projections have been placed within a margin and a title block has been added.

Printing/Plotting

Hard copy (prints or plots on paper) from a variety of different types of AutoCAD drawings of 3D models can be obtained. Some examples have already been shown on p. 71 in Chapter 15.

First example – Printing/plotting (Fig. 19.8)

If an attempt is made to print a multiple viewport screen with all viewport drawings appearing in the plot, only the current viewport will be printed. To print or plot all viewports proceed as folllows:

1. Open a 4-viewport screen of the assembled 3D model shown in the first example (p. 362).

CHAPTER 19

2. Make a new layer **vports** of colour yellow. Make this layer current.
3. *Click* the **Layout1** button in the status bar (Fig. 19.6). The screen changes to a **PSpace** layout.

Fig. 19.6 First example – the **Layout1** button in the status bar

4. The **Paper Space** layout appears with the current viewport outlined in yellow (the colour of the **vports** layer). Using the **Erase** tool, erase the viewports with a *click* on its boundary line. The viewport and its contents disappear.
5. At the command line:

```
Command: enter mv (Mview) right-click
MVIEW
Specify corner of viewport or [ON/OFF/Shadeplot/
  Lock/Object/Polygonal/Restore/2/3/4]: enter 4
  right-click
Specify first corner of viewport or [Fit]: <Fit>:
  right-click
Regenerating model.
Command:
```

and four viewports reappear with the 3D model drawing in each.

6. *Click* the **Model** button to return to **Model Space**. With a *click* in each viewport in turn and using the **ViewCube** settings set viewports in **Front**, **Right**, **Top** and **Isometric** views respectively. Return the screen to **Paper Space**.
7. Turn the layer **vports** off with a *click* on its **Turn a layer On or Off** icon.
8. *Click* the **Plot** tool icon in the **Output/Plot** panel (Fig. 19.7). A **Plot** dialog appears.

Fig. 19.7 The **Plot...** tool icon from the **Output/Plot** panel

9. Check that the printer/plotter is correct and the paper size is also correct.
10. *Click* the **Preview** button. The full preview of the plot appears (Fig. 19.8).

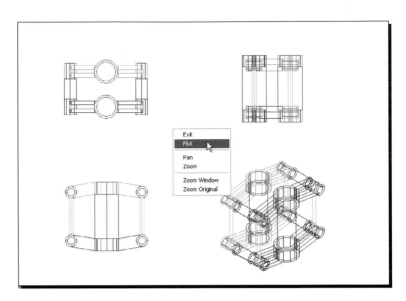

Fig. 19.8 First example – printing/plotting

11. *Right-click* anywhere in the drawing and *click* on **Plot** in the right-click menu which then appears.

12. The drawing plots (or prints).

Second example – Printing/plotting (Fig. 19.9)

1. Open the orthographic drawing with its raster image (Fig. 19.5).

2. While still in **Model Space** *click* the **Plot** tool icon. The **Plot** dialog appears. Check that the required printer/plotter and paper size have been chosen.

3. *Click* the **Preview** button.

4. If satisfied with the preview (Fig. 19.9), *right-click* and in the menu which appears *click* the name **Plot**. The drawing plots.

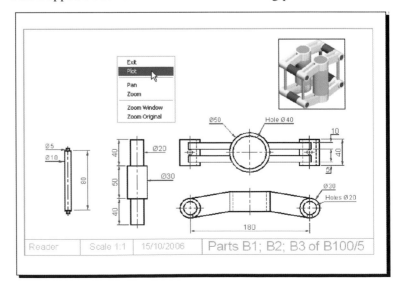

Fig. 19.9 Second example – printing/plotting

Third example – Printing/plotting (Fig. 19.10)

1. Open the 3D model drawing of the assembly shown in Fig. 19.8 in a single **ViewCube/Isometric** view.
2. While in **MSpace**, *click the* **Plot** tool icon. The **Plot** dialog appears.
3. Check that the plotter device and sheet sizes are correct. *Click* the **Preview** button.
4. If satisfied with the preview (Fig. 19.10) *right-click* and *click* on **Plot** in the menu which appears. The drawing plots.

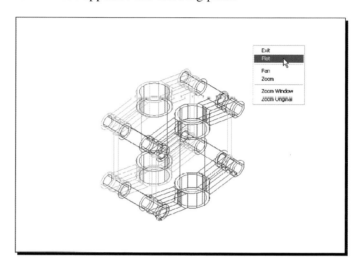

Fig. 19.10 Third example – printing/plotting

Polygonal viewports (Fig. 19.11)

This example illustrates the construction of polygonal viewports is based upon Exercise 6 (p. 370). When the 3D model for this assignment has been completed in **Model Space** proceed as follows:

1. Make a new layer **Yellow** of colour yellow and make this layer current.
2. *Click* the **Layout1** button in the status bar.
3. Erase the viewport with a *click* on its bounding line. The outline and its contents are erased.
4. At the command line:

```
Command: enter mv right-click
[prompts]: enter 4 right-click
[prompts]: right-click
Regenerating model.
Command:
```

and the model appears in a 4-viewport layout.

5. *Click* the **Model** button. With a *click* in each viewport in turn and using the **ViewCube** settings set viewports in **Front**, **Right**, **Top** and **Isometric** views respectively.

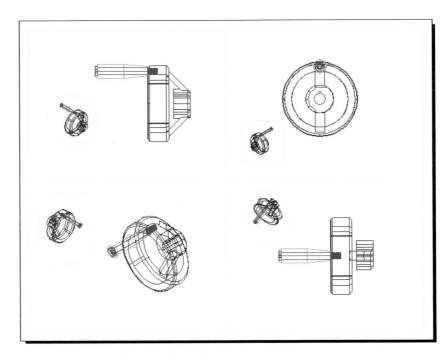

Fig. 19.11 Polygonal viewports – plot preview

  6. **Zoom** each viewport to **All**.
  7. *Click* the **Layout1** button to turn back to **PSpace**.
  8. *Enter* **mv** at the command line, which shows:

```
Command: enter mv right-click
MVIEW
[prompts]: enter p (Polygonal) right-click
Specify start point: In the top right viewport
  pick one corner of a square
Specify next point or [Arc/Close/Length/Undo]:
  pick next corner for the square
Specify next point or [Arc/Close/Length/Undo]:
  pick next corner for the square
Specify next point or [Arc/Close/Length/Undo]:
  enter c (Close) right-click
Regenerating model.
Command:
```

and a square viewport outline appears in the top right viewport within which is a copy of the model.

  9. Repeat in each of the viewports with different shapes of polygonal viewport outlines (Fig. 19.11).
10. *Click* the **Model** button.

11. In each of the polygonal viewports make a different isometric view. In the bottom right viewport change the view using the **3D Orbit** tool.
12. Turn the layer **Yellow** off. The viewport borders disappear.
13. *Click* the **Plot** icon. Make plot settings in the **Plot** dialog.
    Click on the **Preview** button of the **Plot** dialog. The **Preview** appears (Fig. 19.12).

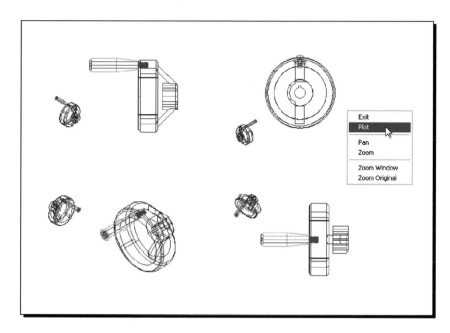

Fig. 19.12 Polygonal viewports – plot preview after turning the layer **Yellow** off

Exercises

Methods of constructing answers to the following exercises can be found in the free website:

http://books.elsevier.com/companions/9780750689830

1. Working to the shapes and sizes given in Fig. 19.13, construct an assembled 3D model drawing of the spindle in its two holders, add lighting and apply suitable materials and render (Fig. 19.14).

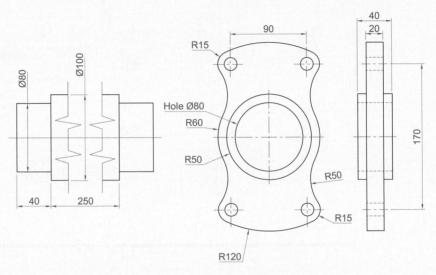

Fig. 19.13 Exercise 1 – details of shapes and sizes

Fig. 19.14 Exercise 1

2.   Figure 19.15 shows a rendering of the model for this exercise and Fig. 19.16 an orthographic projection giving shapes and sizes for the model. Construct the 3D model, add lighting, apply suitable materials and render.

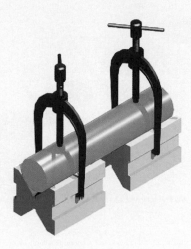

Fig. 19.15 Exercise 2

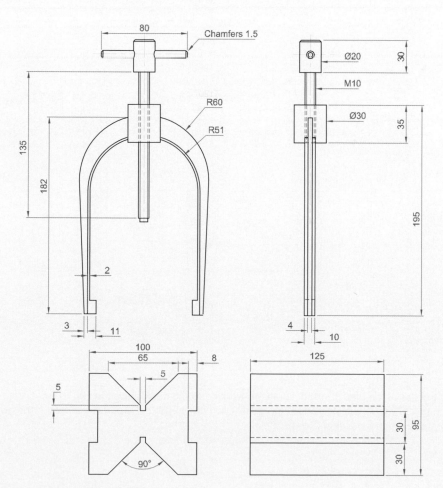

Fig. 19.16 Exercise 2 – orthographic projection

3. Construct a 3D model drawing to the details given in Fig. 19.18. Add suitable lighting and apply a material; then render as shown in Fig. 19.17.

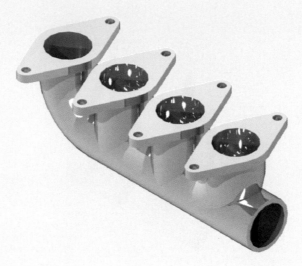

Fig. 19.17 Exercise 3 – isometric drawing

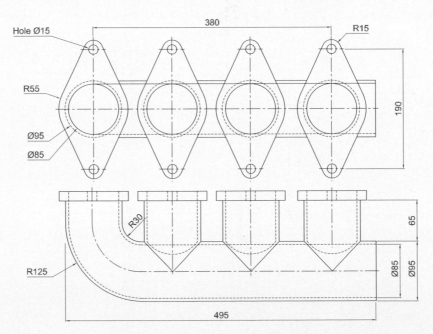

Fig. 19.18 Exercise 3

4. Construct an assembled 3D model drawing working to the details given in Fig. 19.19.
When the 3D model drawing has been constructed disassemble the parts as shown in the given exploded isometric drawing (Fig. 19.20).

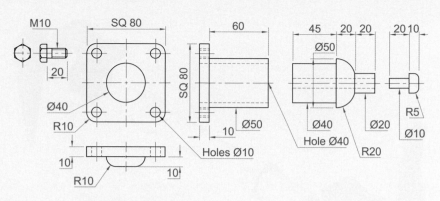

Fig. 19.19 Exercise 4 – details of shapes and sizes

Fig. 19.20 Exercise 4 – exploded rendered model

5.   Working to the details shown in Fig. 19.21, construct an assembled 3D model, with the parts in their correct positions relative to each other. Then separate the parts as shown in the 3D rendered model drawing shown in Fig. 19.22. When the 3D model is complete add suitable lighting and materials and render the result.

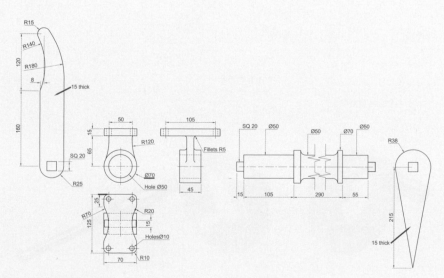

Fig. 19.21 Exercise 5 – details drawing

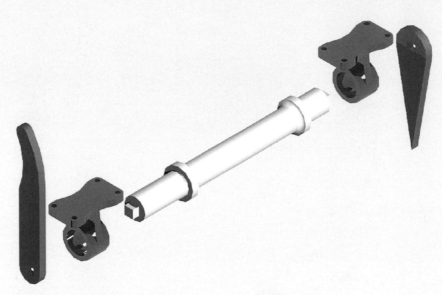

Fig. 19.22 Exercise 5 – exploded rendered drawing

6.   Working to the details shown in Fig. 19.23 construct a 3D model of the parts of the wheel with its handle. Two renderings of 3D models of the rotating handle are shown in Fig. 19.24, one with its parts assembled, the other with the parts in an exploded position relative to each other.

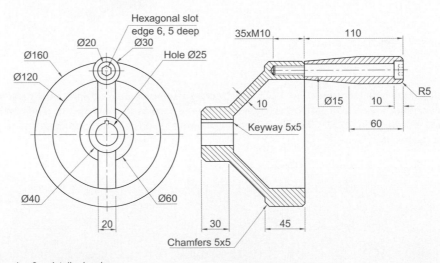

Fig. 19.23 Exercise 6 – details drawing

Fig. 19.24 Exercise 6 – renderings

Part 3

Internet tools and Design

Chapter 20

Internet tools and Help

AIM OF THIS CHAPTER

The aim of this chapter is to introduce the tools which are available in AutoCAD 2009 which make use of facilities available on the World Wide Web (www).

Emailing drawings

As with any other files which are composed of data, AutoCAD drawings can be sent by email as attachments. If security of the drawings is a concern they can be encrypted with a password as the drawings are saved prior to being attached in an email. To encrypt a drawing with a password, *click* **Tools** in the **Save Drawing As** dialog and from the pop-up list which appears *click* **Security Options…** (Fig. 20.1). Then in the **Security Options** dialog which appears (Fig. 20.2) *enter* a password in the **Password or phrase to open this drawing** field, followed by a *click* on the **OK** button. After *entering* a password *click* the **OK** button and *enter* the password in the **Confirm Password** dialog which appears.

Fig. 20.1 Selecting **Security Options…** in the **Save Drawing As** dialog

Fig. 20.2 **Entering** and confirming a password in the **Security Options** dialog

Fig. 20.3 The **Password** dialog appearing when a password encrypted drawing is about to be opened

The drawing then cannot be opened until the password is *entered* in the **Password** dialog (Fig. 20.3) which appears when an attempt is made to open the drawing by the person receiving the email.

There are many reasons why drawings may require to be password-encrypted in order to protect confidentiality of the contents of drawings.

Creating a web page (Fig. 20.7)

To create a web page which includes AutoCAD drawings *left-click* **Publish to Web** in the **Output/Publish** panel (Fig. 20.4).

Fig. 20.4 The **Publish to Web** icon in the **Output/Publish** panel

A series of **Publish to Web** dialogs appear, some of which are shown here in Figs 20.5 to 20.7. After making entries in the dialogs which come on screen after each **Next** button is *clicked*, the resulting web page such as that shown in Fig. 20.7 will be seen (which, in the dialog **Enable i-drop**, can be posted on the web). A *double-click* in any of the thumbnail views in this web page and another page appears, showing the selected drawing in full.

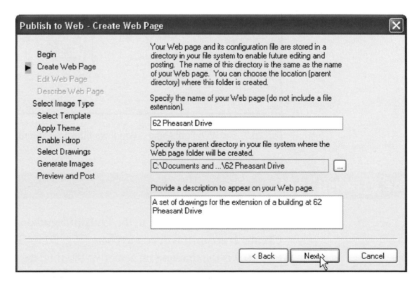

Fig. 20.5 The **Publish to Web – Create Web Page** dialog

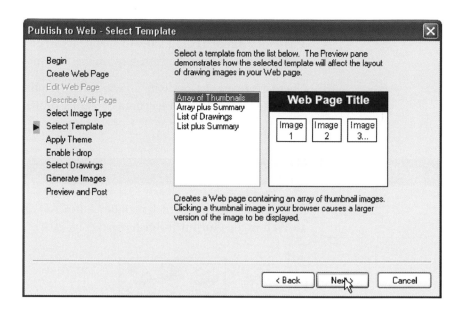

Fig. 20.6 The **Publish to Web – Select Template** dialog

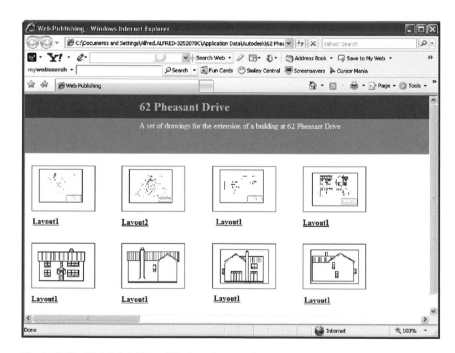

Fig. 20.7 The **Web Publishing – Windows Internet Explorer** page

The eTransit tool

Click the **eTransmit** tool in the **Output/Send** panel (Fig. 20.8) and the **Create Transmittal** dialog appears (Fig. 20.9). The transmittal shown in Fig. 20.9 is the drawing on screen at the time the transmittal was made. Fill in details as necessary. *Click* the **Transmittal Setups…** button and fill in details as necessary in the dialogs which follow. The transmittal is sent as a **zip** file.

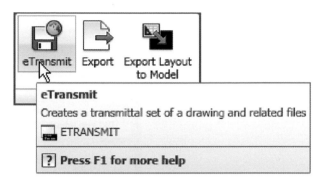

Fig. 20.8 The **eTransmit** icon in the **Output/Send** panel

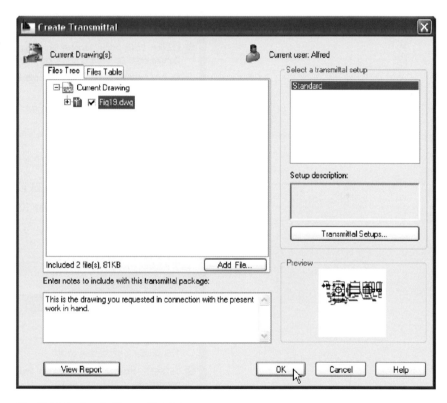

Fig. 20.9 The **Create Transmittal** dialog

A zip file is easier and quicker to email than a drawing file. The AutoCAD drawing can be obtained by unzipping the **zip** file at the receiving end.

Help

Figure 20.10 shows a method of getting help. In this example help on using the **eTransmit** tool is required. *Enter* **eTransmit** in the **InfoCenter** field, followed by a *click* on the **InfoCenter** button. A list appears from

which the operator can select what he/she considers to be the most
appropriate response.

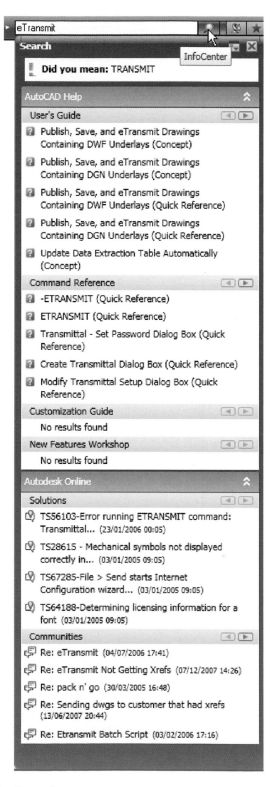

Fig. 20.10 Help for **eTransmit**

The **New Features Workshop**

Enter **New Features Workshop** in the **InfoCenter** field and select **New Features Workshop** from the list which appears. Figure 20.11 shows one of the animations from the workshop describing how to use the **View Cube**.

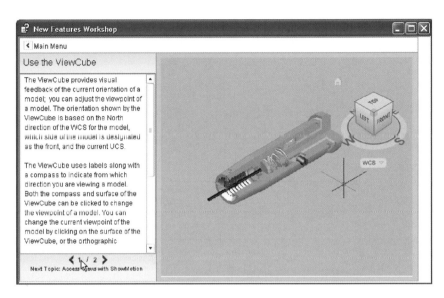

Fig. 20.11 The **New Features Workshop** showing the method of using the **View Cube**

Design and AutoCAD 2009

10 Reasons for using AutoCAD

1. A CAD software package such as AutoCAD 2009 can be used to produce any form of technical drawing.
2. Technical drawings can be produced much more speedily using AutoCAD than when working manually – probably as much as 10 times as quickly when used by skilled AutoCAD operators.
3. Drawing with AutoCAD is less tedious than drawing by hand – features such as hatching, lettering and adding notes are easier, quicker and indeed more accurate to construct.
4. Drawings or parts of drawings can be moved, copied, scaled, rotated, mirrored and inserted into other drawings without having to redraw.
5. AutoCAD drawings can be saved to a file system without necessarily having to print the drawing. This can save the need for large paper drawing storage areas.
6. The same drawing or part of a drawing need never be drawn twice, because it can be copied or inserted into other drawings with ease. A basic rule when working with AutoCAD is: **Never draw the same feature twice**.
7. New details can be added to drawings or be changed within drawings without having to mechanically erase the old details.
8. Dimensions can be added to drawings with accuracy, reducing the possibility of making errors.
9. Drawings can be plotted or printed to any scale without having to redraw.
10. Drawings can be exchanged between computers and/or emailed around the world without having to physically send the drawing.

The place of AutoCAD 2009 in designing

The contents of this book are only designed to help those who have limited (or no) knowledge and skills of the construction of technical drawings using AutoCAD 2009. However, it needs to be recognized that the impact of modern computing on the methods of designing in industry has been immense. Such features as analysis of stresses, shear forces and bending forces can be carried out more quickly and accurately using computing methods. The storage of data connected with a design and the recovery of the data speedily are carried out much more easily using computing methods than prior to the introduction of computing.

AutoCAD 2009 can play an important part in the design process because technical drawings of all types are necessary for achieving a well-designed artefact whether it be an engineering component, a machine, a building, an electronics circuit or any other design project.

In particular, two-dimensional drawings (2D drawings) that can be constructed in AutoCAD 2009 are still of great value in modern industry.

AutoCAD 2009 can also be used to produce excellent and accurate three-dimensional (3D) models, which can be rendered to produce photographic-like images of a suggested design. Although not dealt with in this book, data from 3D models constructed in AutoCAD 2009 can be taken for use in computer-aided machining (CAM).

At all stages in the design process, either (or both) 2D or 3D drawings play an important part in aiding those engaged in design to assist in assessing the results of their work at various stages. It is in the design process that drawings constructed in AutoCAD 2009 play an important part.

In the simplified design process chart shown in Fig. 21.1 an asterisk (*) has been shown against those features where the use of AutoCAD 2009 can be regarded as being of value.

A design chart (Fig. 21.1)

The simplified design chart (Fig. 21.1) shows the following features.

- **Design brief:** a design brief is a necessary feature of the design process. It can be in the form of a statement, but it is usually much more. A design brief can be a written report which not only includes a statement made of the problem which the design is assumed to be solving, but includes preliminary notes and drawings describing difficulties which

Fig. 21.1 A simplified design chart

may be encountered in solving the design and may include charts, drawings, costings etc., to emphasize some of the needs in solving the problem for which the design is being made.

- **Research:** the need to research the various problems which may arise when designing is often much more demanding than the chart (Fig. 21.1) shows. For example the materials being used may require extensive research as to costing, stress analysis, electrical conductivity, difficulties in machining or in constructional techniques and other such features.
- **Ideas for solving the brief:** this is where technical and other drawings and sketches play an important part in designing. It is only after research that designers can ensure the brief will be fulfilled.
- **Models:** these may be constructed models in materials representing the actual materials which have been chosen for the design, but in addition 3D solid model drawings, such as those which can be constructed in AutoCAD 2009, can be of value. Some models may also be made in the materials from which the final design is to be made so as to allow testing of the materials in the design situation.
- **Chosen solution:** this is where the use of drawings constructed in AutoCAD 2009 is of great value. 2D and 3D drawings come into their own here. It is from such drawings that the final design will be manufactured.
- **Realization:** the design is made. There may be a need to manufacture a number of the designs in order to enable evaluation of the design to be fully assessed.
- **Evaluation:** The manufactured design is tested in situations where it is liable to be placed in use. Evaluation will include reports and notes which could include drawings with suggestions for amendments to the working drawings from which the design was realized.

Enhancements in AutoCAD 2009

AutoCAD 2009 contains major enhancements over previous releases, whether working in a **2D** or a **3D** workspace. Please note that not all the enhancements in AutoCAD 2009 are described in this introductory book. Among the more important enhancements are the following:

1. The most obvious enhancement is when AutoCAD 2009 is loaded. The AutoCAD application window is very different from previous versions.
2. Some dialogs changed to palettes – e.g. the **Layer Properties Manager** is now a palette.
3. The introduction of the **Ribbon** replacing the **DASHBOARD**.
4. The **Ribbon** includes **6 Tabs** – **Home**, **Blocks & References**, **Annotate Tools**, **View** and **Output**.
5. When a tab is selected with a *click*, the selected tab shows a number of panels. Each **Panel** contains a number of tools icons.

6. In the **3D Modeling** workspace the introduction of the **ViewCube** enabling speedy changes to viewing planes.
7. Changes to methods of constructing sheet sets.
8. New commands added.
9. Enhancements in dimensioning.
10. Enhancements in **Multiple Text**.
11. A wide variety of methods of manipulating layers.
12. New **Help** features in an **InfoCenter** (top right-hand corner of AutoCAD 2009 window).
13. Enhancements in methods of lighting and illuminating 3D models when rendering.
14. Enhancements in multileaders.
15. Some new set variables.

Annotation scaling

Annotation scaling has many uses. This example shows only one of its many uses.

1. Figure 21.2 shows a scale **1:30** front view of a bungalow in a **Paper Space** window in the **AutoCAD Classic** workspace.
2. *Click* the down-pointing arrow in the right-hand end of the status bar and, in the menu which appears, note the drawing has been constructed to scale of **1:30** (Fig. 21.2).
3. *Click* **1:50** in the Annotation Scaling menu and the front view changes to show a scale **1:50** front view (Fig. 21.3).

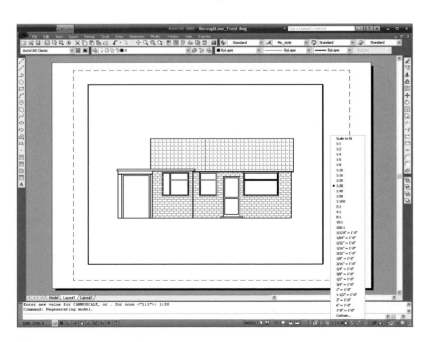

Fig. 21.2 First example – annotation scaling. The front view of a bungalow drawn to a scale of 1:30

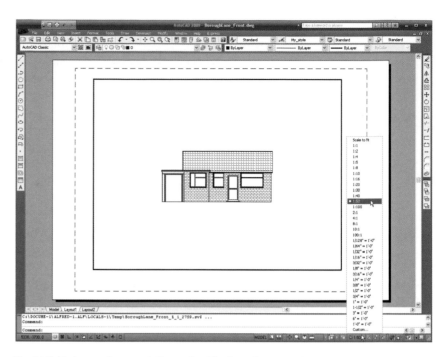

Fig. 21.3 First example – annotation scaling. The front views scaled to 1:50

Multileaders

There are a variety of uses for multileaders. The second example shows one such use. The drawing in which multileaders are to be included is shown in Fig. 21.4.

1. In the **Annotate/Multileaders** panel *click* the **Multileader Style** icon (Fig. 21.5). The **Multileader Style Manager** dialog appears. *Click* its **Modify...** button.

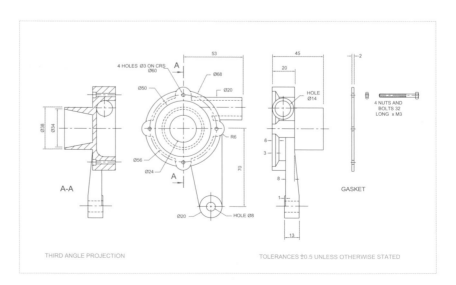

Fig. 21.4 Second example – multileaders. The drawing to which multileaders are to be added

Fig. 21.5 Second example – **Multileader Style** icon in the **Annotate/Multileaders** panel

2. In the **Modify Multileaders Style Standard** dialog which appears, make entries as shown in Fig. 21.6.
3. *Click* the **Multileader** icon in the panel.
4. In the drawing, in response to prompts at the command line, add multileaders one after the other to the drawing shown in Fig. 21.4, numbering the parts **1** to **4**. See Fig. 21.7.

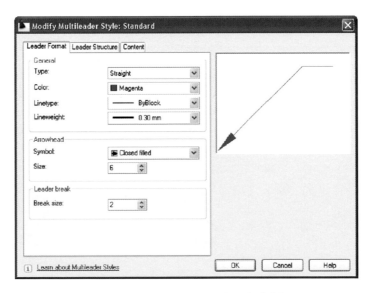

Fig. 21.6 Second example – **Modify Multileaders Style Standard** dialog

System requirements for running AutoCAD 2009

- **Note:** there are two editions of AutoCAD 2009 – 32 bit and 64 bit editions.
- **Operating system:** Windows XP Professional, Windows XP Professional (x64 Edition), Windows XP Home Edition, Windows 2000 or Windows Vista 32 bit, Windows Vista 64 bit.
- **Microsoft Internet Explorer 7.0**.
- **Processor:** Pentium III 800 Mhz.
- **Ram:** at least 128 MB.

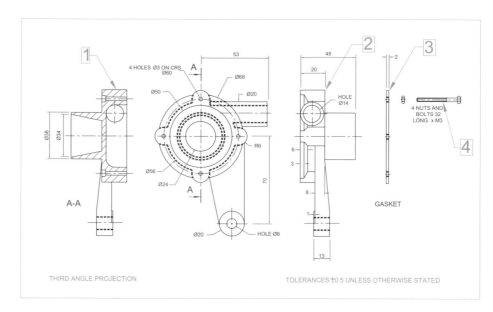

Fig. 21.7 Second example – multileaders

- **Monitor screen:** 1024×768 VGA with True Colour as a minimum.
- **Hard disk:** a minimum of 300 MB.
- **Graphics card:** an AutoCAD certified graphics card. Details can be found on the web page **AutoCAD Certified Hardware XML Database**.

Part 4

Appendices

Appendix A

Printing/plotting

Introduction

Some suggestions for printing/plotting of AutoCAD drawings have already been given (pp. 297–299). Plotters or printers can be selected from a wide range of different types and are used for printing or plotting drawings constructed in AutoCAD 2009. The example given here is from a print using one of the default printers connected to the computer that the author was using. However, if another plotter or printer is connected to the computer, its driver can be set by first opening the Windows **Control Panel**, and with a *double-click* on the **Autodesk Plotter Manager** icon the **Plotters** dialogs appear (Fig. A.1).

Fig. A.1 The **Plotters** window

Double-click on the **Add, remove or change plotter properties** icon and the **Plotter Configuration Editor** dialog appears. Add settings as required in the first of the dialogs from this editor (Fig. A.2). There are several more dialogs in the series in which selections will need to be made before completing the setting up of a printer or plotter for the printing of AutoCAD drawings.

Note

1. AutoCAD drawings can be printed from the default printers already installed in the Windows system of the computer on which AutoCAD 2009 is loaded.

2. Plots or prints from drawings constructed in AutoCAD 2009 can be made from either **Model Space** or from **Paper Space**.

Fig. A2   The first of the series of **Plotter Configuration Editor** dialogs

An example of a printout

1. Either select **Plot** with a *click* on icon in the **Output/Plot** panel (Fig. A.3) or from the **File** drop-down menu. The **Plot** dialog appears (Fig. A.4)

Fig. A3   The **Plot** tool icon in the **Output/Plot** panel

2. There are two parts to the **Plot** dialog. Figure A4 shows both parts. A *click* on the arrow at the bottom-right-hand corner of the dialog closes to reveal only the left-hand part and vice versa.

3. Select an appropriate printer or plotter from the **Printer/Plotter** list. In this example this is a colour printer. Then select the correct paper size from the **Paper Size** pop-up list. Next, select what is to be printed/plotted from the **What to Plot** pop-up list – in the example shown this is **Display**. *Click* the **Properties** button and in the **Custom Properties** dialog (not shown) set orientation to **Landscape** (dot in radio button). Then *click* the **Preview** button of the **Plot** dialog.

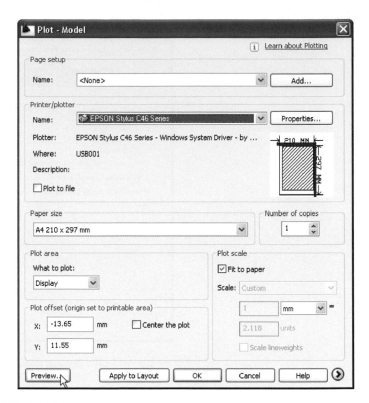

Fig. A4 The **Plot** dialog

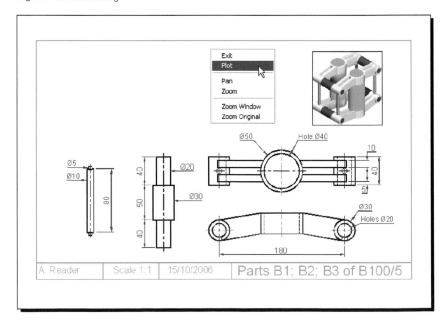

Fig. A.5 The **Plot Preview** window with its *right-click* menu

4. A preview of the drawing to be printed/plotted appears (Fig. A.5). If satisfied with the preview, *right-click* and from the menu which appears *click* **Plot**. If not satisfied *click* **Exit**. The preview disappears and the **Plot** dialog reappears. Make changes as required from an inspection of the preview and carry on in this manner until a plot can be made.

Appendix B

List of tools

Introduction

AutoCAD 2009 allows the use of over 300 tools. Some operators prefer using the word 'commands', although command as an alternative to tool is not in common use today. The majority of these tools are described in this list. Many of the tools described here have not been used in this book, because it is an introductory text designed to initiate readers into the basic methods of using AutoCAD 2009. It is hoped the list will encourage readers to experiment with those tools not described in the book. The abbreviations for tools which can be abbreviated are included in parentheses after the tool name. Tool names can be *entered* in upper or lower case.

A list of 2D tools is followed by a listing of 3D tools. Internet tools are described at the end of this listing. Not all of the tools available in AutoCAD 2009 are shown here.

2D tools

About – Brings the **About AutoCAD** bitmap on screen

Adcenter (dc) – Brings the **DesignCenter** palette on screen

Align (al) – Aligns objects between chosen points

Appload – Brings the **Load/Unload Applications** dialog to screen

Arc (a) – Creates an arc

Area – States in square units of the area selected from a number of points

Array (ar) – Creates **Rectangular** or **Polar** arrays in 2D

Ase – Brings the **dbConnect Manager** on screen

Attdef – Brings the **Attribute Definition** dialog on screen

Attedit – Allows editing of attributes from the Command line

Audit – Checks and fixes any errors in a drawing

Autopublish – Creates a **DWF** file for the drawing on screen

Bhatch (h) – Brings the **Boundary Hatch** dialog on screen

Block – Brings the **Block Definition** dialog on screen

Bmake (b) – Brings the **Block Definition** dialog on screen

Bmpout – Brings the **Create Raster File** dialog on screen

Boundary (bo) – Brings the **Boundary Creation** dialog on screen

Break (br) – Breaks an object into parts

Cal – Calculates mathematical expressions

Chamfer (cha) – Creates a chamfer between two entities

Chprop (ch) – Brings the **Properties** window on screen

Circle (c) – Creates a circle

Copy (co) – Creates single or multiple copies of selected entities

Copyclip (Ctrl+C) – Copies a drawing, or part of a drawing for inserting into a document from another application

Copylink – Forms a link between an AutoCAD drawing and its appearance in another application such as a word processing package

Copytolayer – Copies objects from one layer to another

Customize – Brings the **Customize** dialog to screen, allowing the customization of toolbars, palettes etc.

Dashboard – Has the same action as **Ribbon**

Dashboardclose – Closes the **Ribbon**

Ddattdef (at) – Brings the **Attribute Definition** dialog to screen

Ddatte (ate) – Edits individual attribute values

Ddcolor (col) – Brings the **Select Color** dialog on screen

Ddedit (ed) – The **Text Formatting** dialog box appears on selecting text

Ddim (d) – Brings the **Dimension Style Manager** dialog box on screen

Ddinsert (i) – Brings the **Insert** dialog on screen

Ddmodify – Brings the **Properties** window on screen

Ddosnap (os) – Brings the **Drafting Settings** dialog on screen

Ddptype – Brings the **Point Style** dialog on screen

Ddrmodes (rm) – Brings the **Drafting Settings** dialog on screen

Ddunits (un) – Brings the **Drawing Units** dialog on screen

Ddview (v) – Brings the **View Manager** on screen

Del – Allows a file (or any file) to be deleted

Dgnexport – Creates a **MicroStation V8 dgn** file from the drawing on screen

Dgnimport – Allows a **MicroStation V8 dgn** file to be imported as an AutoCAD dwg file

Dim – Starts a session of dimensioning

Dim1 – Allows the addition of a single addition of a dimension to a drawing

Dimension tools – The **Dimension** toolbar contains the following tools: **Linear**, **Aligned**, **Arc Length**, **Ordinate**, **Radius**, **Jogged**, **Diameter**, **Angular**, **Quick Dimension**, **Baseline**, **Continue**, **Quick Leader**, **Tolerance**, **Center Mark**, **Dimension Edit**, **Dimension Edit Text**, **Update** and **Dimension Style**

Dist (di) – Measures the distance between two points in coordinate units

Distantlight – Creates a distant light

Divide (div) – Divides and entity into equal parts

Donut (do) – Creates a donut

Dsviewer – Brings the **Aerial View** window on screen

Dtext (dt) – Creates dynamic text. Text appears in drawing area as it is *entered*

Dxbin – Brings the **Select DXB File** dialog on screen

Dxfin – Brings the **Select File** dialog on screen

Dxfout – Brings the **Save Drawing As** dialog on screen

Ellipse (el) – Creates an ellipse

Erase (e) – Erases selected entities from a drawing

Exit – Ends a drawing session and closes AutoCAD 2009

Explode (x) – Explodes a block or group into its various entities

Explorer – Brings the Windows **Explorer** on screen

Export (exp) – Brings the **Export Data** dialog on screen

Extend (ex) – To extend an entity to another

Fillet (f) – Creates a fillet between two entities

Filter – Brings the **Object Selection Filters** dialog on screen

Gradient – Brings the **Hatch and Gradient** dialog on screen

Group (g) – Brings the **Object Grouping** dialog on screen

Hatch – Allows hatching by the *entry* responses to prompts

Hatchedit (he) – Allows editing of associative hatching

Help – Brings the **AutoCAD 2009 Help: User Documentation** dialog on screen

Hide (hi) – To hide hidden lines in 3D models

Id – Identifies a point on screen in coordinate units

Iimageattach (iat) – Brings the **Select Image File** dialog on screen

Iinsertobj – Brings the **Insert Object** dialog on screen

Imageadjust (iad) – Allows adjustment of images

Imageclip – Allows clipping of images

Import – Brings the **Import File** dialog on screen

Insert (i) – Brings the **Inert** dialog on screen

Isoplane (Ctrl/E) – Sets the isoplane when constructing an isometric drawing

Join (j) – Joins lines which are in line with each other or arcs which are from the same centre point

Laycur – Changes layer of selected objects to current layer

Laydel – Deletes and purges a layer with its contents

Layer (la) – Brings the **Layer Properties Manager** dialog on screen

Layout – Allows editing of layouts

Lengthen (len) – Lengthens an entity on screen

Limits – Sets the drawing limits in coordinate units

Line (l) – Creates a line

Linetype (lt) – Brings the **Linetype Manager** dialog on screen

List (li) – Lists in a text window details of any entity or group of entities selected

Load – Brings the **Select Shape File** dialog on screen

Ltscale (lts) – Allows the linetype scale to be adjusted

Measure (me) – Allows measured intervals to be placed along entities

Menu – Brings the **Select Customization File** dialog on screen

Menuload – Brings the **Load/Unload Customizations** dialog on screen

Mirror (mi) – Creates an identical mirror image to selected entities

Mledit – Brings the **Multiline Edit Tools** dialog on screen

Mline (ml) – Creates mlines

Mlstyle – Brings the **Multiline Styles** dialog on screen

Move (m) – Allows selected entities to be moved

Mslide – Brings the **Create Slide File** dialog on screen

Mspace (ms) – When in PSpace changes to MSpace

Mtext (mt or t) – Brings the **Multiline Text Editor** on screen

Mview (mv) – To make settings of viewports in Paper Space

Mvsetup – Allows drawing specifications to be set up

New (Ctrl+N) – Brings the **Select template** dialog on screen

Notepad – For editing files from the Windows **Notepad**

Offset (o) – Offsets selected entity by a stated distance

Oops – Cancels the effect of using **Erase**

Open – Brings the **Select File** dialog on screen

Options – Brings the **Options** dialog to screen

Ortho – Allows ortho to be set ON/OFF

Osnap (os) – Brings the **Drafting Settings** dialog to screen

Pagesetup – Brings either the **Page Setup Manager** on screen

Pan (p) – Drags a drawing in any direction

Pbrush – Brings Windows **Paint** on screen

Pedit (pe) – Allows editing of polylines. One of the options is **Multiple** allowing continuous editing of polylines without closing the command

Pline (pl) – Creates a polyline

Plot (Ctrl+P) – Brings the **Plot** dialog to screen

Point (po) – Allows a point to be placed on screen

Polygon (pol) – Creates a polygon

Polyline (pl) – Creates a polyline

Preferences (pr) – Brings the **Options** dialog on screen

Preview (pre) – Brings the print/plot preview box on screen

Properties – Brings the **Properties** palette on screen

Psfill – Allows polylines to be filled with patterns

Psout – Brings the **Create Postscript File** dialog on screen

Purge (pu) – Purges unwanted data from a drawing before saving to file

Qsave – Saves the drawing file to its current name in AutoCAD 2009

Quickcalc (qc) – Brings the **QUICKCALC** palette to screen

Quit – Ends a drawing session and closes down AutoCAD 2009

Ray – A construction line from a point

Recover – Brings the **Select File** dialog on screen to allow recovery of selected drawings as necessary

Recoverall – Repairs damaged drawing

Rectang (rec) – Creates a pline rectangle

Redefine – If an AutoCAD command name has been turned off by **Undefine**, **Redefine** turns the command name back on

Redo – Cancels the last **Undo**

Redraw (r) – Redraws the contents of the AutoCAD 2009 drawing area

Redrawall (ra) – Redraws the whole of a drawing

Regen (re) – Regenerates the contents of the AutoCAD 2009 drawing area

Regenall (rea) – Regenerates the whole of a drawing

Region (reg) – Creates a region from an area within a boundary

Rename (ren) – Brings the **Rename** dialog on screen

Revcloud – Forms a cloud-like outline around objects in a drawing to which attention needs to be drawn

Ribbon – Brings the ribbon on screen

Ribbonclose – Closes the ribbon

Save (Ctrl+S) – Brings the **Save Drawing As** dialog box on screen

Saveas – Brings the **Save Drawing As** dialog box on screen

Saveimg – Brings the **Render Output File** dialog on screen

Scale (sc) – Allows selected entities to be scaled in size – smaller or larger

Script (scr) – Brings the **Select Script File** dialog on screen

Setvar (set) – Can be used to bring a list of the settings of set variables into an AutoCAD Text window

Shape – Inserts an already loaded shape into a drawing

Shell – Allows MS-DOS commands to be *entered*

Sketch – Allows freehand sketching

Solid (so) – Creates a filled outline in triangular parts

Spell (sp) – Brings the **Check Spelling** dialog on screen

Spline (spl) – Creates a spline curve through selected points

Splinedit (spe) – Allows the editing of a spline curve

Status – Shows the status (particularly memory use) in a Text window

Stretch (s) – Allows selected entities to be stretched

Style (st) – Brings the **Text Styles** dialog on screen

Tablet (ta) – Allows a tablet to be used with a pointing device

Tbconfig – Brings the **Customize** dialog on screen to allow configuration of a toolbar

Text – Allows text from the Command line to be *entered* into a drawing

Thickness (th) – Sets the thickness for the Elevation command

Tilemode – Allows settings to enable Paper Space

Tolerance – Brings the **Geometric Tolerance** dialog on screen

Toolbar (to) – Brings the **Customize User Interface** dialog on screen

Trim (tr) – Allows entities to be trimmed up to other entities

Type – Types the contents of a named file to screen

UCS – Allows selection of **UCS** (user Coordinate System) facilites

Undefine – Suppresses an AutoCAD command name

Undo (u) (Ctrl+Z) – Undoes the last action of a tool

View – Brings the **View** dialog on screen

Vplayer – Controls the visibility of layers in Paperspace

Vports – Brings the **Viewports** dialog on screen

Vslide – Brings the **Select Slide File** dialog on screen

Wblock (w) – Brings the **Create Drawing File** dialog on screen

Wipeout – forms a polygonal outline within which all crossed parts of objects are erased

Wmfin – Brings the **Import WMF** dialog on screen

Wmfopts – Brings the **WMF in Options** dialog on screen

Wmfout – Brings the **Create WMF File** dialog on screen

Xattach (xa) – Brings the **Select Reference File** dialog on screen

Xline – Creates a construction line

Xref (xr) – Brings the **Xref Manager** dialog on screen

Zoom (z) – Brings the zoom tool into action

3D tools

3Darray – Creates an array of 3D models in 3D space

3Dcorbit – Allows methods of manipulating 3D models on screen

3Ddistance – Allows the controlling of the distance of 3D models from the operator

3Ddwf – Brings up the **Export 3D DWF** dialog

3Dface (3f) – Creates a 3 or 4 sided 3D mesh behind which other features can be hidden

3Dfly – Allows walkthroughs in any 3D plane

3Dforbit – controls the viewing of 3D models without constraint

3Dmesh – Creates a 3D mesh in 3D space

3Dmove – Shows a 3D move icon

3Dorbit (3do) – Allows a continuous movement and other methods of manipulation of 3D models on screen

3Dorbitctr – Allows further and a variety of other methods of manipulation of 3D models on screen

3Dpan – Allows the panning of 3D models vertically and horizontally on screen

3Drotate – Displays a 3D rotate icon

3Dsin – Brings the **3D Studio File Import** dialog on screen

3Dsout – Brings the **3D Studio Output File** dialog on screen

3Dwalk – Starts walk mode in 3D

Align – Allows selected entities to be aligned to selected points in 3D space

Ameconvert – Converts AME solid models (from Release 12) into AutoCAD 2000 solid models

Anipath – Opens the **Motion Path Animation** dialog

Box – Creates a 3D solid box

Cone – Creates a 3D model of a cone

Convertoldlights – Converts lighting from previous releases to AutoCAD 2009 lighting

Convertoldmaterials – Converts materials from previous releases to AutoCAD 2009 materials

Convtosolid – Converts plines and circles with thickness to 3D solids

Convtosurface – Converts objects to surfaces

Cylinder – Creates a 3D cylinder

Dducs (uc) – Brings the **UCS** dialog on screen

Edgesurf – Creates a 3D mesh surface from four adjoining edges

Extrude (ext) – Extrudes a closed polyline

Flatshot – Brings the **Flatshot** dialog to screen

Freepoint – Point light created without settings

Freespot – Spot light created without settings

Helix – Construct a helix

Interfere – Creates an interference solid from selection of several solids

Intersect (in) – Creates an intersection solid from a group of solids

Light – Enables different forms of lighting to be placed in a scene

Lightlist – Opens the **Lights in Model** palette

Loft – Activates the **Loft** command

Materials – Opens the **Materials** palette

Matlib – Outdated instruction

Mirror3d – Mirrors 3D models in 3D space in selected directions

Mview (mv) – When in PSpace brings in MSpace objects

Pface – Allows the construction of a 3D mesh through a number of selected vertices

Plan – Allows a drawing in 3D space to be seen in plan (UCS World)

Planesurf – Creates a planar surface

Pointlight – Allows a point light to be created

Pspace (ps) – Changes MSpace to PSpace

Pyramid – Creates a pyramid

-render – can be used to make rendering settings from the command line. Note the hyphen (-) must precede **render**

Renderpresets – Opens the **Render Presets Manager** dialog

Renderwin – Opens the **Render** window

Revolve (rev) – Forms a solid of revolution from outlines

Revsurf – Creates a solid of revolution from a pline

Rmat – Brings the **Materials** palette on screen

Rpref (rpr) – Opens the **Advanced Render Settings** palette

Section (sec) – Creates a section plane in a 3D model

Shade (sha) – Shades a selected 3D model

Slice (sl) – Allows a 3D model to be cut into several parts

Solprof – Creates a profile from a 3D solid model drawing

Sphere – Creates a 3D solid model sphere

Spotlight – Creates a spotlight

Stlout – Saves a 3D model drawing in ASCII or binary format

Sunproperties – Opens the **Sun Properties** palette

Sweep – Creates a 3D model from a 2D outline along a path

Tabsurf – Creates a 3D solid from an outline and a direction vector

Torus (tor) – Allows a 3D torus to be created

Ucs – Allows settings of the UCS plane

Ucs – Allows settings of the UCS plane

Union (uni) – Unites 3D solids into a single solid

View – Creates view settings for 3D models

Visualstyles – Opens the **Visual Styles Manager** palette

Vpoint – Allows viewing positions to be set from x,y,z entries

Vports – Brings the **Viewports** dialog on screen

Wedge (we) – Creates a 3D solid in the shape of a wedge

Xedges – Creates a 3D wireframe for a 3D solid

Internet tools

Etransmit – Brings the **Create Transmittal** dialog to screen

Publish – Brings the **Publish** dialog to screen

Appendix C
Some set variables

Introduction

AutoCAD 2009 is controlled by a large number of variables (over 460 in number), the settings of many of which are determined when making entries in dialogs. Others have to be set at the command line. Some are read-only variables which depend upon the configuration of AutoCAD 2009 when it is originally loaded into a computer (default values). Only a limited number of the variables are shown here.

A list of those set variables follows which are of interest in that they often require setting by *entering* figures or letters at the command line. To set a variable, *enter* its name at the command line and respond to the prompts which arise.

To see all set variables, *enter* **set** (or **setvar**) at the command line:

```
Command: enter set right-click
SETVAR Enter variable name or ?: enter ?
Enter variable name to list ⊂*⊃ : right-click
```

and a **Text** window opens showing a first window with a list of the first of the variables. To continue with the list press the **Return** key when prompted and at each press of the **Return** key another window opens.

To see the settings for each set variable *enter* the name of the variable at the command line, followed by pressing the **F1** key which brings up the **Help** screen; *click* the search tab, followed by *entering* set variables in the **Ask** field. From the list then displayed the various settings of all set variables can be read.

Some of the set variables

ANGDIR – Sets angle direction. **0** counterclockwise; **1** clockwise

APERTURE – Sets size of pick box in pixels

AUTODWFPUBLISH – Sets **Autopublish** on or off

BLIPMODE – Set to **1** marker blips show; set to **0** no blips

COMMANDLINE – Opens the command line palette.

COMMANDLINEHIDE – Closes the command line palette

COPYMODE – Sets whether **Copy** repeats

> **Note**
>
> **DIM** variables – There are over 70 variables for setting dimensioning, but most are in any case set in the **Dimension Styles** dialog or as dimensioning proceeds. However, one series of the **Dim** variables may be of interest:
>
> **DMBLOCK** – Sets a name for the block drawn for an operator's own arrowheads. These are drawn in unit sizes and saved as required
>
> **DIMBLK1** – Operator's arrowhead for first end of line
>
> **DIMBLK2** – Operator's arrowhead for other end of line

DRAGMODE– Set to **0** no dragging; set to **1** dragging on; set to **2** automatic dragging

DRAG1 – Sets regeneration drag sampling. Initial value is 10

DRAG2 – Sets fast dragging regeneration rate. Initial value is 25

FILEDIA – Set to **0** disables **Open** and **Save As** dialogs; set to **1** enables these dialogs

FILLMODE – Set to **0** hatched areas are filled with hatching. Set to **0** hatched areas are not filled. Set to **0** and plines are not filled.

GRIPS – set to **1** and grips show. Set to **0** and grips do not show.

LIGHTINGUNITS – set to **1** (international) or **2** (USA) for photometric lighting to function

MBUTTONPAN – Set to **0** on *right-click* menu with the Intellimouse. Set to **1** Intellimouse *right-click* menu on

MIRRTEXT – Set to **0** text direction is retained; set to **1** text is mirrored

NAVVCUBE – Sets the **ViewCube** on/off

NAVVCUBELOCATION – Controls the position of th **ViewCube** between top-right (**0**) and bottom-left (**3**)

NAVVCUBEOPACITY – Controls the opacity of the **ViewCube** from **0** (hidden) to **100** (dark)

NAVVCUBESIZE – controls the size of the **ViewCube** between **0** (small) to **2** (large)

PELLIPSE – Set to **0** creates true ellipses; set to **1** polyline ellipses

PERSPECTIVE – Set to **0** places the drawing area into parallel projection. Set to **1** places the drawing area into perspective projection

PICKBOX – Sets selection pick box height in pixels

PICKDRAG – Set to **0** selection windows picked by two corners; set to **1** selection windows are dragged from corner to corner

RASTERPREVIEW – set to **0**, raster preview images not created with drawing. Set to **1**, preview image created

SHORTCUTMENU – For controlling how *right-click* menus show:

0 all disabled; **1** default menus only; **2** edit mode menus; **4** command mode menus; **8** command mode menus when options are currently available. Adding the figures enables more than one option.

SURFTAB1 – Sets mesh density in the M direction for surfaces generated by the **Surfaces** tools

SURFTAB2 – Sets mesh density in the N direction for surfaces generated by the **Surfaces** tools

TEXTFILL – Set to **0**, True Type text shows as outlines only; set to **1**, True Type text is filled

TiILEMODE – Set to **0** Paperspace enabled; set to **1** tiled viewports in Modelspace

TOOLTIPS – Set to **0** no tool tips; set to **1** tool tips enabled

TPSTATE – Set to **0** and the Tool Palettes window is inactive. Set to **1** and the Tool Palettes window is active

TRIMMODE – Set to **0** edges not trimmed when **Chamfer** and **Fillet** are used; set to **1** edges are trimmed

UCSFOLLOW – Set to **0** new UCS settings do not take effect; set to **1** UCS settings follow requested settings

UCSICON – Set **OFF** UCS icon does not show; set to **ON** it shows

Index